THE ESTIMATION
OF PROBABILITIES

An Essay on Modern Bayesian Methods

THE ESTIMATION
OF PROBABILITIES
An Essay on Modern Bayesian Methods

IRVING JOHN GOOD

RESEARCH MONOGRAPH NO. 30
THE M.I.T. PRESS, CAMBRIDGE, MASSACHUSETTS

First M.I.T. Press Paperback Edition, August 1968.

Library of Congress Catalog Card Number: 65–13192

ISBN 978-0-262-57015-2 (pb : alk. paper)

This monograph is
dedicated to
William Ernest Johnson
the teacher of
John Maynard Keynes
and
Harold Jeffreys

Foreword

There has long been a need in science and engineering for systematic publication of research studies larger in scope than a journal article but less ambitious than a finished book. Much valuable work of this kind is now published only in a semiprivate way, perhaps as a laboratory report, and so may not find its proper place in the literature of the field. The present contribution is the thirtieth of the M.I.T. Press Research Monographs, which we hope will make selected timely and important research studies readily accessible to libraries and to the independent worker.

J. A. STRATTON

Preface

The estimation of probabilities is of practical and philosophical interest, and can be difficult when the sample is small, difficult enough in fact so that the literature is not as extensive as one might expect, although some of it is ancient. This monograph contains a review of much of the immediately relevant literature known to me, but the main emphasis is on methods that are new or have not been written up in a connected manner.

The difficulties become clear when it is realized that we estimate probabilities every minute of the day, at least implicitly, and that how we do this is unknown. When this problem is solved, a potential pathway to artificial intelligence will be cleared, apart from easier applications, such as to character recognition and medical diagnosis. The work described in this monograph is only a fraction of all that is required for these purposes. A more complete treatment would involve much discussion of problems of classification. For this purpose, another monograph will be necessary.

We shall be concerned with methods for the estimation of probabilities from "effectively small" samples, and with some implications of these methods for tests of significance for multinomial samples and for contingency tables. Most of the techniques described depend on a modern Bayesian approach. The method mentioned in Chapter 8, for multinomial sampling when the number of categories is large, as in the sampling of species or of vocabulary, originated in a suggestion made by A. M. Turing in 1941, but was not published until 1953. Herbert

Robbins pointed out, when I described the method in a recent lecture, that it is an example of what he now calls the "empirical Bayes method." The earlier chapters make much use of a hierarchy of three types of probabilities, usually to be roughly interpreted as physical, logical, and subjective, although I have called them Types I, II, and III in order to decrease controversy. (They *could* all be physical and they *could* all be subjective.) In my opinion, this is a necessary improvement of the Bayesian methods that are usually used. In Chapter 9 the emphasis shifts to the maximum-entropy principle and multidimensional contingency tables.

Fuller abstracts are given in the Introduction (Chapter 1) and Summary (Chapter 10).

It is a pleasure to acknowledge the facilities provided by the Communications Research Division of the Institute for Defense Analyses, and especially the expert mathematical typing of Mrs. Euthie Anthony, the use of the electronic computer, and the help of Mr. G. J. Mitchell in extracting bugs from my FORTRAN programs. I am also much indebted to Dr. Oscar Rothaus and Dr. Lloyd Welch for literally invaluable help in connection with Appendices D and E, respectively.

Princeton, New Jersey   IRVING JOHN GOOD
April, 1964

Contents

THE ESTIMATION
OF PROBABILITIES

An Essay on Modern Bayesian Methods

1. Introduction

There are at least four reasons for being interested in the estimation of probabilities:

1. For betting, including insurance decisions and for all other decisions and predictions.
2. For discrimination between hypotheses, when these hypotheses are not simple statistical hypotheses. (When the hypotheses are "simple," the probabilities of relevant events have uncontroversial values.) This second reason can be understood in the light of examples mentioned later.
3. For deciding whether an event is surprising. If it is surprising, the correctness of the observations should be checked.
4. For the improved understanding of the philosophy of probability and statistics—and of the nature of probability judgments.

The basic theme in the following pages is the estimation of probabilities from *effectively small samples*. By "effectively small," we mean that the sample is small for the purpose of estimating some probability, though the sample might be absolutely large. The problem is not quite the same as that of estimating small probabilities, in which part of the technique is to enlarge the sample. (See, for example, Haldane[47, 48] and Pearson.[90]) We shall suppose for the most part that the sample already available is to be used by itself for the estimation. Circumstances are always changing, but humans have the facility of estimating the probabilities of many events that have never previously occurred. Their

3

sample is their past experience, and they must often make predictions without the benefit of an increased sample. Suppose, for example, that we wish to estimate the probability that there will be a storm here tomorrow. It might be suggested that the probability could be taken as the proportion of days in February on which there are storms, or of those days on which there are storms when there has been no storm for 24 hours and the temperature at noon was 53 degrees and at 2:30 P.M. was 51.7 degrees. The totality of information is such that, if it were all used, the proportion would be of the indeterminate form 0/0; this would be true even if we ignored all the information that we judged to be virtually irrelevant or, at any rate, if this were done in any obvious manner. This difficulty would remain even if we could give a precise definition of a storm. The difficulty is well known to actuaries.

Nevertheless, for the purpose of making decisions, we do manage to make approximate estimates of probabilities. How this is done is an interesting problem in psychology and in neurophysiology. It might, for example, be conjectured that neural circuits automatically use a maximum-entropy estimation (see Chapter 9). The problem of estimating probabilities of events that have never occurred is philosophically interesting and is, in my opinion, likely to be important for the design of ultraintelligent machines. The unconscious goal of the scientific philosopher is the automation of science.

A completely general solution of the problem of estimating probabilities seems entirely out of reach, and this book therefore deals only with a number of simple problems: binomial and multinomial estimation, multinomial sampling when the number of categories is very large (the "sampling of species"), estimation of small probabilities in a large pure contingency table, and in a multidimensional population contingency table.

The estimation of probabilities leads directly to the discussion of other matters, such as tests for no association in a contingency table when the sample is not necessarily large, estimates of the "coverage" of a sample of species or words, and the inconsistency of maximum-likelihood estimation with a Bayesian philosophy.

Each distinct problem of probability estimation is liable to need somewhat different methods, but there are several unifying themes in this monograph. In particular, we rely mainly on a modern Bayesian approach. Some of the methods involve the notion of a Type III distribution (in a sense described in Chapter 2, not in Karl Pearson's sense!). Sometimes the initial distributions have parameters in them, and can be regarded as "semi-initial" after the parameters are estimated. The method mentioned in Chapter 8 is Bayesian but uncontroversial. There is some discussion of the principle of maximum entropy for generating null hypotheses, and of invariance theories. Included is the use of the device of imaginary results, which involves Bayes' theorem in reverse, of methods that seem to be *good enough*, and of methods that *improve* on previous methods; since final answers do not seem to be attainable even in the simplest of problems. My own view, following Keynes and Koopman, is that judgments of probability inequalities are possible but not judgments of exact probabilities; therefore a Bayesian should have upper and lower betting probabilities. In order to avoid complications, this viewpoint is de-emphasized in this monograph. We stress at this point, however, that a convenient approximate method of using this philosophy is to make use of two or more initial distributions, in effect, as several alternative models. This will give rise to a number of final betting probabilities, the largest and smallest of which would be the upper and lower betting probabilities.

In all honesty, the Bayesian will wish to ascribe different weights to different initial (or Type II) distributions. These weights will again typically lie in intervals of values, but, for simplicity, they can be taken to be precise numbers. They then act as a Type III distribution, but the weighted average of the set of Type II distributions, weighted with the Type III distribution, is in effect a single Type II distribution. Thus the Bayesian can use the Type III distribution merely as an aid to his own judgment, and is not compelled to state the Type III distribution aloud. But to do so helps to explain what otherwise might seem to be a peculiar and unnatural-looking Type II distribution. The main example of this method in this book is in connection with initial Type II distributions that are of the symmetrical Dirichlet form.

An alternative to the use of a Type III distribution is to plot a graph or make a table of the Type II likelihoods. This is an example of a compromise between Bayesian and non-Bayesian methods, since one of the techniques used by the non-Bayesian is to graph the ordinary (Type I) likelihoods rather than making a weighted combination of them as in the usual Bayesian method.

If a unique initial distribution (a "credibility" distribution) could be generally agreed, for any class of estimation problems, then there would be no need to make use of distributions of Type III. Such agreement has not yet been reached, and I do not believe that it ever will be. (Compare Good,[27] p. 48.)

Although this book is largely a survey, most of it has not previously been published; it has seemed to me to be more valuable to state old results and prove new ones rather than to describe in detail what is already conveniently available.

The reader might find it useful to read the Summary (Chapter 10) before proceeding to Chapter 2.

2. Kinds of Probability

When two people mean different things by the same words, they are often said to have different philosophies, and controversy arises. The word "probability" is used in several different ways, and I shall try to minimize controversy by careful choice of terminology. (See Good.[27, 28, 36]) For example, a *physical probability* (also called "material probability," "intrinsic probability," "propensity," or "chance") is a probability that is regarded as an intrinsic property of the material world, existing irrespective of minds and logic. Many people believe in the existence of such probabilities, and it is interesting to talk about them even if we deny their existence. (I am inclined to believe in them myself.) A similar qualification applies to all other kinds of probability discussed herein. A *psychological probability* is a degree of belief or intensity of conviction that is used for betting purposes, for making decisions, or for any other purpose, not necessarily after mature consideration and not necessarily with any attempt at "consistency" with one's other opinions. These have been studied, for example, by John Cohen[12] and by Davidson, Suppes, and Siegel.[15] When a person or persons, called "you," uses a fairly *consistent* set of probabilities, they are called *subjective* ("personal") or *multisubjective* ("multipersonal") probabilities. Thus a subjective probability is a special case of a psychological one, but psychological probabilities, especially those of very young children, are very often *blatantly* inconsistent. A *consistent* set of probabilities is one against which it is not possible to make a "Dutch book," that is, a selection of bets by an opponent against which you are bound to lose money whatever happens. It has been proved (for example, Ramsey,[95] von Neumann and Morgenstern,[89] Savage,[100] and Smith[104]) that a consistent set of probabilities obeys the usual axioms of probability, except that *Kolmogorov's axiom* (complete additivity) is inessential.

Ramsey seems to have been the first to have deduced axioms of subjective probability and utility from a set of compelling desiderata, but he did not think it worthwhile to carry the argument "to several decimal places," perhaps because he underestimated the magnitude of the opposition.

A *logical probability*, or, in the terminology of F. Y. Edgeworth[19] and Bertrand Russell,[99] a "credibility," is a rational intensity of conviction, implicit in the given information, and such that if a person does not agree with it he is wrong. Jeffreys[63] and Keynes[71] formulated theories in which credibilities are central; but, in his biography of Ramsey,[72] Keynes nobly recanted from this position and admitted that subjective probabilities are primary and that credibilities might not exist. Savage,[100] a subjectivist, calls credibilities "necessary probabilities," but does not believe in their existence. Keynes argued that credibilities are only partially ordered; Jeffreys states in the preface of his book on probability[63] that *this* is what Keynes withdrew, but this was a slip. *Intuitive probability* is *either* subjective or logical probability. Koopman,[74, 75] without reference to utilities, which are used in most other justifications, showed that some compelling but complicated desiderata, concerning partially ordered intuitive probabilities, imply that these probabilities can be *represented* by numbers, satisfying the usual axioms, and such that the partial ordering is not contradicted by these representations. In Good,[27] I argued that subjective probabilities are partially ordered, and described what seemed to me then, and still seems, to be the simplest theory that one can use in order to apply partially ordered subjective probabilities to ordinary statistical problems. A similar theory for utilities is implicit therein but is more explicit in Good.[28, 30, 38] Savage[100] put more emphasis on decisions from the start. Both de Finetti[21] and Savage[100] apparently regard subjective probability as the *only* kind of probability worth the name; my own position is less extreme, for example, I regard it as *mentally healthy* to believe that credibilities exist and as subjectively probable that physical probabilities exist. It is sometimes possible to compromise between subjective probability and credibility, see, for example, Good,[39, 40] in fact perhaps most applications of either are compromises between the two.

A *tautological* or purely mathematical probability is one that is introduced by *definition*. For example, a "simple statistical hypothesis"

H is, by definition, a proposition such that, for various other propositions or events E, the probability $P(E \mid H)$ has an assigned numerical value, which is therefore tautological. The statement, "H is *true*," implies that $P(E \mid H)$ equals the physical probability of E. I regard this as a *linguistic axiom*, or in more technical terms, an axiom of the metalanguage.

Further historical note: Poisson[94] emphasized very much the distinction between physical and nonphysical probabilities. Ramsey[95] said there might well be two kinds of probability, and Bartlett[2] also was, and perhaps still is, a "dualist." (Quantum mechanics encourages philosophical dualism, in its present form.†) Jeffreys uses the word "chance" or "intrinsic probability" for a *physical probability* and is somewhat noncommittal about its existence. In the current century, Carnap has greatly emphasized the distinction again. The need for further classification was pointed out by Kemble,[69] and a classification into at least five kinds was given in some detail by Good.[36]

2.1 Bayesians

Several different kinds of Bayesians exist, but it seems to me that the essential defining property of a Bayesian is that *he regards it as meaningful to talk about the probability $P(H \mid E)$ of a hypothesis H, given evidence E*. Consequently, he will make more use of Bayes' theorem than a non-Bayesian will. Bayes' theorem itself is a trivial consequence of the product axiom of probability, and it is not a belief in this theorem that makes a person a Bayesian. Rather it is a readiness to incorporate intuitive probability into statistical theory and practice, and into the philosophy of science and of the behavior of humans, animals, and automata, and in an understanding of all forms of communication, and everything.

The *mathematics* used by a Bayesian can be interpreted without agreeing with his philosophy, since one can always imagine models in which hypotheses have physical probabilities. A familiar method of doing this is to imagine several urns, each containing balls of several colors. Each urn corresponds to one hypothesis. An ordinary sampling experiment is analogous to the selection of balls from a single urn, usually with replacement and shuffling. But, before taking a sample from an urn, we can select an urn at random, too. In such a situation, the probabilities occurring in the two kinds of sampling can both be

---

† This is denied by A. Landé, *From Dualism to Unity in Quantum Physics*, Cambridge University Press (1960), but see E. Wigner, "The Problem of Measurement," *Amer. J. Phys.*, **31**, 6–15, especially p. 6 (1963).

regarded as physical probabilities. But, even when these probabilities are of the same philosophical kind, it is convenient to refer to the sampling from the urns as "sampling of Type I," and to the sampling of an urn from the collection of urns as "sampling of Type II." We can define sampling of Types III, IV, and so forth, by means of an obvious extension of the model. This terminology of a hierarchy of types is natural and will later be extended to some other analogous self-explanatory uses, but the "principles of rationality of Types I and II," in Good,[38] involve a somewhat different meaning of the word "type."

We can imagine ordinary (Type I) sampling from populations, each of which can itself be obtained by (Type II) sampling from a super-population, or population of Type II, whose elements are ordinary populations (of Type I). The probabilities that govern the ordinary sampling are here called probabilities of Type I, and those governing the selection of populations from the superpopulation as probabilities of Type II. (See Good[28] and Good,[29] p. 862.) We use this terminology *whether or not* the Type II probabilities are physical. With this under-standing, the terminology of "Type I" and "Type II" probabilities is philosophically neutral. We might appear to be assuming that the Type I probabilities are necessarily physical, but an extreme subjectivist need not be worried by this since he would presumably accept de Finetti's definition of physical probability in terms of subjective probability. This will be explained in the next chapter.

Although controversy can be avoided in the mathematics of proba-bility, it is unavoidable when the mathematics is applied to the outside world. It will then be found, in many applications, that the Type II probabilities are necessarily intuitive. Even in such cases, Type III probabilities are in my opinion of some interest: a Type III probability might be the subjective probability that some credibility or credibility distribution belongs to some specified set. (See Good.[28]) Examples are given in later chapters.

Suppose that you wish to estimate a physical probability p on the basis of some evidence E. A Bayesian will assume an *initial* ("prior") Type II probability distribution for p. (I am here ignoring partial ordering of intuitive probabilities in order to avoid complications—it would need to be allowed for in a complete discussion.) The evidence E converts the distribution into a *final* ("posterior") Type II distribu-tion $F(p)$. You might then take the expectation

$$\hat{p} = \mathscr{E}_{\mathrm{II}}(p \mid E) = \int_0^1 p \, dF(p) \qquad (2.1)$$

as your point estimate of p, and you might also give interval estimates and calculate the Type II variance of p,

$$\text{var}_{\text{II}}(p \mid E) = \int_0^1 (p - \hat{p})^2 \, dF(p) \qquad (2.2)$$

For the purpose of betting, only the Type II expectation $\hat{p}$ is relevant, although some people have argued that the whole distribution, especially the Type II variance, should be relevant. (See, for example, Allais.[1]) Many people would feel happier making bets concerning the throws of a symmetrical-looking die if they had taken a large sample first, but *logical* symmetry of the information concerning the faces is really enough for the assertion of their Type II equiprobability *on the next throw*. It is important to realize that the estimates need recalculation after each throw, since the result of each throw changes the available information. Some absurd criticisms of the Bayesian philosophy are made through overlooking this point. It is also important, when judging a betting probability, to allow for the possibility that you are in a "game situation," that is, that an opponent might have had some influence over the selection or construction of the die. Even if you are in a game situation, then the Type II expectation of p is equal to your Type II probability of a "success" on the next "trial." But of course, your beliefs concerning the opponent's strategy will influence the expectation.

2.2 Compromises Between Bayesian and Non-Bayesian Methods

An *extreme* Bayesian believes that every intuitive probability is *precise*, whereas less extreme Bayesians regard intuitive probabilities as only partially ordered so that each probability merely lies in some interval of values. In view of Koopman's results (see also C. A. B. Smith[104]), the less extreme Bayesian can assume that the probabilities are precise and satisfy the usual numerical axioms, but he holds in mind that the set of all probabilities is not at all unique. He makes judgments of probability inequalities and infers new probability inequalities with the help of the mathematical theory. He also explicitly recognizes that there is in each application an optimal amount of self-interrogation that is worthwhile. (See Good.[38]) This necessarily leads to a compromise between the Bayesian and non-Bayesian positions; indeed, if the interval associated with each intuitive probability is taken as (0, 1), then the Bayesian is converted into a non-Bayesian. One is *more or less* a Bayesian depending on the precision with which one is prepared to make intuitive probability estimates.

Just as the non-Bayesian finds it expedient to construct mathematical models of Type I probability distributions in which he tries to minimize the number of parameters, the Bayesian will do the same, but with both the Type I and the Type II probability distributions. This leads to Type II problems of estimation and significance. Examples will be given in later chapters.

Whether or not the subjectivist is extreme, he should be interested in the arguments and opinions of non-Bayesians, of credibilists, and of other Bayesians, since everything is grist to the mill of his own judgments.

Another example of a compromise between Bayesian and non-Bayesian methods is as follows. (Further examples are mentioned in Chapter 5.) Consider any statistical technique with which you have some sympathy. Find out whether it is equivalent to the use of an initial probability distribution by using Bayes' theorem in reverse. (See Good,[27] p. 81.) *Then replace this initial distribution by a better one.*

As an example, consider the method of confidential prediction used by Thatcher[106] for binomial sampling. It is equivalent to the use of *two* initial (divergent) densities p^{-1} and $(1 - p)^{-1}$. If we want only one initial distribution, and that a symmetrical one, it is natural to average these two distributions. The result is Haldane's[47] divergent initial density $p^{-1}(1 - p)^{-1}$. (Or see Jeffreys,[63] p. 123.) But this distribution leads to maximum-likelihood estimation of p, which is unsatisfactory (see Chapter 3 and page 36). It can be improved upon by using the more general symmetrical beta distribution, or a weighted combination of such distributions.

Likewise, what A. D. Roy[98] calls "pistimetric inference" for multinominal distributions can be seen at once to be equivalent to the use of the initial density proportional to $\prod p_i^{-1}$ where p_i is the physical probability of the ith category. This distribution again leads to maximum-likelihood estimation and so can be improved by allowing any symmetrical initial Dirichlet distribution. Pistimetric inference is very similar to fiducial inference. Fiducial inference also is sometimes equivalent to the use of an initial distribution, but it can lead to inconsistencies with Bayesian inference. It too can perhaps be improved. (See Appendix A.)

3. Simple Sampling

A *simple sample* results from a sequence of *trials* each of which can be either a *success* or a *failure*, denoted by A and B, respectively, or by 1 and 0. Thus a simple sample is a sample from a space or *alphabet* consisting of two letters. The term "alphabet" or "generalized alphabet" is often used in information theory and has the advantage of sounding more concrete than "sampling space." The result of any finite sample is a finite binary sequence. The length of the sequence is the *sample size*.

A binomial sample is a simple sample that results in a *random* binary sequence, that is, a sequence in which the successive trials are statistically independent. If the sample size is given as N, then every sequence having just r successes and s failures, where $r + s = N$, is equiprobable, and has probability

$$p^r(1 - p)^s$$

where p is the probability of a successful trial.

When p is unknown, it might have been sampled from a super-population or Type II population and have a Type II initial distribution $F(.)$; that is, the Type II initial probability that $p \leq x$ is

$$P_{II}(p \leq x) = F(x)$$

When this is so, the Type II probability of each sequence which has r successes and s failures, when the sample size N is given, is

$$\int_0^1 x^r(1 - x)^s \, dF(x)$$

12

and the total Type II probability that there will be r successes and s failures is of course obtained by multiplying by the binomial coefficient $\binom{N}{r}$.

A converse theorem, of considerable interest for the foundations of probability, was first proved by de Finetti[21] and will shortly be stated.

Let us say that a stochastic binary sequence is *permutable* if, whenever r and s are given, the probability $Q(r, s)$ that a segment of the process of length $r + s$ will have r successes and s failures at specified places in the segment is independent of where these places are. This is the *permutability postulate* of the philosopher William Ernest Johnson.[66] Such a sequence is said to be "symmetric" by Savage,[100] or "equivalent" by de Finetti.[21] In our application of the notion of a permutable sequence, we shall interpret the probability just mentioned as a Type II probability. The simple sampling can itself be reasonably described as "permutable." De Finetti's theorem can be stated in the following words:

THEOREM 3.1

A simple sample that generates a permutable binary sequence can always be regarded as having been obtained by binomial sampling in which the Type I probability p has a Type II initial distribution. Moreover this Type II distribution is unique. In other words $Q(r, s)$ necessarily has a representation of the form

$$Q(r, s) = \int_0^1 p^r (1 - p)^s \, dF(p) \tag{3.1}$$

where $F(.)$ is a unique distribution function. In short, permutable simple sampling is binomial sampling with an unknown value of p.†

Before proving this theorem, let us consider de Finetti's philosophical interpretation. It is that if you take a simple sample and accept the permutability postulate, then your subjective probabilities $Q(r, s)$, if consistent, will be the same as if you had assumed an initial subjective probability distribution for a parameter p, which parameter you can then *define* as the "true physical probability" if you feel inclined. His point of view then, shared by Savage,[100] is that subjective probability enables one to define physical probability and statistical independence as mathematical fictions. I do not personally take so extreme a point of view, since it seems to me that one would not accept the permutability

---

† See Fréchet[24] and Hewitt and Savage[56] for extensions and further references. They say that "Haag seems to have been the first author to discuss symmetric sequences …," but he was perhaps anticipated by W. E. Johnson.[66]

postulate unless one already had the notion of physical probability and (approximate) statistical independence at the back of one's mind. Similarly, for long sequences, any particular simple result, such as alternation, 010101010101010101, would seem to me initially more probable than any particular complicated result, both because a lack of statistical independence could not be ruled out (a physical reason) and because of the simplicity (an intuitive reason).

Although my philosophy differs from that of de Finetti and Savage on this point, I agree with them in regarding subjective probability as necessary for the purpose of measuring physical probability, and as an extension of logic that is necessary for science in general and for statistics and decision-making.

Proof:†

Here $Q(., .)$ must satisfy the consistency condition

$$Q(r + 1, s) + Q(r, s + 1) = Q(r, s) \tag{3.2}$$

Let

$$\Delta Q(r, s) = Q(r, s) - Q(r + 1, s) = Q(r, s + 1)$$

Then, for any positive integer q,

$$\Delta^q Q(r, 0) = \Delta^{q-1} Q(r, 1) = \cdots = Q(r, q) \geq 0 \tag{3.2a}$$

Therefore the sequence $\{Q(r, 0)\}$, where $r = 0, 1, 2, \cdots$, is "totally monotonic" ("totally decreasing"): see, for example, G. H. Hardy,[51] p. 253, for the terminology. Therefore, $Q(r, 0)$ is of the form

$$Q(r, 0) = \int_0^1 p^r \, dF(p)$$

where F is an increasing and bounded function of p, by Hausdorff's theorem, Hardy,[51] Theorem 207. Also, F is unique by Hardy's Theorem 203. Moreover $Q(0, 0) = 1$, so that F is a probability distribution function. Formula 3.1 now readily follows from the consistency equation in the form of Equation 3.2a.

---

† *Historical note:* This proof and the note on the Hausdorff method of summation in Appendix B were drawn to the attention of the editor of *Mathematical Tables and Aids to Computation* in 1953 and were generalized to the case of multiple sampling in 1955, when the writer was working under an ONR contract at Princeton University. (See Chapter 4.) Savage,[100] p. 53n, also states that "de Finetti's theorem can be proved very quickly and naturally by applying the theory of the Hausdorff moment problem ...., but this method does not seem to generalize readily." A short proof for simple sampling was also given by Hinčin.[58]

3.1 The Estimation of p

Suppose now that we wish to estimate the physical probability p of a binomial distribution, after we have taken a sample of size N. Let the number of successes be r. We know that our sample sequence is permutable, so that a knowledge of r and N exhausts all the information that the sample can give regarding the value of p. If we are Bayesians, we will believe in the existence of intuitive probabilities $Q(r, s)$ and hence in that of a Type II initial distribution $F(.)$ for p. A non-Bayesian might deny the existence of both $Q(., .)$ and $F(.)$ or would refuse to try to guess them, and he would probably fall back on either confidence intervals or maximum likelihood.

Let us first discuss maximum-likelihood estimation. The maximum-likelihood estimate of p is simply r/N. This is a perfectly good estimate if r and $s = N - r$ are both large. But then we would hardly require any theory. Since we are concerned with probability estimation from effectively small samples, let us consider the case where $r = 0$. The maximum-likelihood estimate of p is then 0. It is not easy to see how this value can be used for betting purposes or for making decisions. If you really believe that a probability is 0, you should be prepared to give arbitrarily large odds against a success on the next trial. Unless N were fabulously large, no one in his senses would be able to make this use of the maximum-likelihood estimate (when $r = 0$). (An exception occurs when a "success" is utterly disastrous; many people would be prepared to give infinite odds against the world's coming to an end because, if they lost, they would not have to settle the account.)

Confidence-interval estimation suffers from a similar disadvantage when r is 0 or small enough; the confidence interval has to include the point $p = 0$. To assert that 0 is a lower bound for p is to assert *nothing at all* about the lower bound and so is not useful for betting purposes or for decision-making. It is easy to be right in a high proportion of cases if one's statements each convey very little information. The theory of confidential estimation would be more satisfactory if it were possible to assign utilities to the estimates. There is usually some utility loss $\psi(S \mid p)$ in stating that a parameter p belongs to some set S, even when this statement is true. This is the sole reason why short intervals are preferred to long ones containing them.

Confidential intervals can also be used for making predictions of the number of successes in a second sample; see the discussion in Chapter 2.

3.2 Bayesian Methods of Estimating p

If $F(.)$, the Type II initial distribution of p, were a known physical probability distribution, the estimation of p from a sample would be uncontroversial. One would simply use Bayes' theorem to obtain a final distribution of p, from which the expectation and other information could be derived mathematically. In some other circumstances the function F might itself be estimable from a sample of samples, but we are mainly interested in the estimation of probabilities, not in the more difficult problem of estimating probability distributions. Accordingly, we assume that we do not have a sample of samples, and the Bayesian must either guess the distribution of F if it is a physical distribution, or he must select F to "represent his ignorance" if it is not. When selecting this Type II distribution, a degree of arbitrariness *seems* impossible to avoid. David Hume[59] has not yet been defeated.

One reason why maximum-likelihood estimation is often preferred to Bayesian estimation is that the maximum-likelihood estimate *looks* less arbitrary. (The *asymptotic* properties of maximum-likelihood estimates are no better than those of any Bayesian methods, and in any case are not very relevant to the problem of estimation from effectively small samples.) The use of maximum likelihood is equivalent to the selection of the *mode* of the final Type II distribution, when the initial distribution is uniform, and the selection of the mode is not the Bayesian procedure. But if the initial Type II density has Haldane's divergent form $p^{-1}(1 - p)^{-1}$, the final expected value of p is the maximum-likelihood estimate. For this reason one must reject Haldane's initial distribution. Haldane also suggested a slight modification, in order to achieve convergence. It leads approximately to maximum-likelihood estimation, so that our objections will still apply, especially when N is small. See also pages 11, 28, 36, and 68, and Appendix E.

When a statistician selects a Type II initial distribution, he ought to make some effort to use the one that corresponds to his own judgment, whether the distribution is physical or intuitive. Since the selection is to some extent arbitrary, the statistician will have an opportunity to cheat. The more honest he tries to be, the more arbitrary and complicated his choice will look, and the more he will open the door to accusations of cheating. There is also the danger of unconscious cheating (wishful thinking). For this reason there is much to be said for minimizing arbitrariness, for compromising between philosophy and politics, between the ideal and the expedient. One method of making such a compromise is to restrict the class of initial (Type II) distributions to a class with a small number of parameters.

A distribution without any parameters at all is the uniform distribution $F(p) = p$ with density function $f(p) = 1$. This selection is often called the "Bayes postulate," not to be confused with Bayes' theorem. It was used by Laplace[78] for our problem of binomial estimation. The inference is known as *Laplace's law of succession* and states that the Type II probability that the next trial will be successful is $(r + 1)/(N + 2)$ and is of course also equal to the Type II expectation of the physical probability. For example, if there is one success in one trial, the probability estimate is 2/3. Although the Bayes postulate has come in for some very severe criticism, it is intuitively obvious that this estimate is better than the maximum-likelihood estimate, which is 1.

Even without a sample, there are occasions when the Bayes postulate is unreasonable. The actuaries, G. F. Hardy[50] and Lidstone,[80] suggested that a more flexible and convenient class of initial densities was given by the beta form, proportional to $p^\alpha(1 - p)^\beta$, where $\alpha > -1$ and $\beta > -1$.

By selecting from the class of beta distributions, the statistician can give expression to his initial ideas concerning both the Type II expectation and the Type II variance of p, and this is about as much flexibility as is likely to be required in some applications. The distribution has only two parameters but covers a good variety of unimodal shapes. Given a sample of r successes in N trials, the final Type II expectation and variance of p are

$$\mathscr{E}_{\mathrm{II}}(p \mid r, s) = \frac{\alpha + r + 1}{\alpha + \beta + N + 2} \tag{3.3}$$

and

$$\mathrm{var}_{\mathrm{II}}(p \mid r, s) = \frac{(\alpha + r + 1)(2\alpha + 2\beta + r + 1)}{(\alpha + \beta + N + 2)(\alpha + \beta + N + 3)} \tag{3.4}$$

The parameters α and β can be determined by equating the guessed initial Type II expectation and variance to

$$\left. \begin{aligned} \mathscr{E}_{\mathrm{II}}(p \mid 0, 0) &= \frac{\alpha + 1}{\alpha + \beta + 2} \\ \mathrm{var}_{\mathrm{II}}(p \mid 0, 0) &= \frac{(\alpha + 1)(2\alpha + 2\beta + 1)}{(\alpha + \beta + 2)(\alpha + \beta + 3)} \end{aligned} \right\} \tag{3.5}$$

Naturally Equation 3.3 reduces to Laplace's formula when $\alpha = \beta = 0$. The more general formula can be expressed as follows: add constants $\alpha + 1$ and $\beta + 1$ to the observed frequencies r and s of successes and failures. In a problem in which the information is symmetrical between successes and failures, then $\alpha = \beta$, and the same constant should be added to both r and s. (Symmetry of the *information* does not imply *physical* symmetry, for which $p = 1/2$ and r and s are infinite.)

It seems possible that G. F. Hardy was the first to suggest a "continuum of inductive methods," to use Carnap's phrase.[9]

The philosopher C. D. Broad[7] extended Laplace's law of succession to the problem of sampling without replacement from a finite population. He found that sampling cannot give a high Type II probability that the whole population is of one kind (all "successful") unless the sample is a large fraction of the whole population. (See Jeffreys,[63] p. 128.) Jeffreys and Wrinch therefore suggested that a nonzero initial Type II probability should be associated with the end points of the unit interval, in order to provide a rationale for scientific induction. Another way of expressing this idea is to say that any natural null hypothesis should be assigned a nonzero Type II probability. If this is not done, the final Type II probability of the hypothesis $p = 0$ remains zero however much evidence there might be in its favor. Actually there are other points on the unit interval that seem to be distinguished, such as the midpoint $p = 1/2$. It seems to me, that, if we adopt the philosophy just mentioned, there should be a nonzero probability associated with every computable value of p, where "computable" is here used in Turing's sense. (See Good,[27] p. 55.) This idea needs modification since we are usually more concerned with whether a hypothesis is *approximately* true. But in this chapter, and in Chapter 4, this aspect of probability estimation will be ignored, and we shall assume that p has a Type II density, unsullied by Dirac delta functions.

If one wished to use a bimodal initial Type II distribution, a reasonable class of densities would be the linear combinations of two beta distributions.

An example of the beta initial Type II density is Haldane's density $p^{-1}(1 - p)^{-1}$, mentioned earlier. In view of Equation 3.3, Haldane's distribution leads to the maximum-likelihood estimate r/N for the final Type II expectation of p.

Jeffreys[63] and Perks[93] independently proposed invariance theories of initial probability distributions for more general problems. Jeffreys' idea was to attribute a total probability to a region of parameter space by a rule that would lead to the same probability for the same region of parameter space when the parameters undergo a change of coordinates. (The parameters here were those of the Type I sampling distribution; in the present application there is one parameter, namely p.) Perks' *basic* method is applicable only to a single parameter. The point of both theories was to try to overcome the important objection to the Bayes postulate, that it depends on the choice of the parameter.

For our present problem, the invariance theories of both Jeffreys and Perks lead to the initial Type II density of the beta form proportional to $p^{-1/2}(1 - p)^{-1/2}$. Hence, by Equation 3.3, they lead to the estimate $(r + \frac{1}{2})/(N + 1)$ for the final Type II probability. This is a compromise between Laplace's estimate and the maximum-likelihood estimate. When there are more than three successes and three failures, there is little difference between the three methods (invariance, Laplace, and maximum-likelihood) for many practical purposes.

If the beta density proportional to $p^\alpha(1-p)^\beta$ is assumed, the estimate of the betting probability that the next m trials will be successful is not the mth power of the estimate $(r + \frac{1}{2})/(N + 1)$. In fact, by the product law of (conditional) probabilities, or by integration, the estimate is

$$\mathscr{E}_{\text{II}}(p^m \mid r, s) = \frac{(r + \alpha + 1)(r + \alpha + 2)\cdots(r + \alpha + m)}{(N + \alpha + \beta + 2)\cdots(N + \alpha + \beta + m + 1)} \quad (3.6)$$

This tends to zero as m tends to infinity, even if $r = N$, because we have neglected to allow a nonzero initial probability to the null hypothesis $p = 1$. (See Jeffreys,[63] p. 128.)

In virtue of de Finetti's theorem, the Jeffreys-Perks law of succession is *equivalent* to the assumption of an initial Type II distribution of the special beta form mentioned earlier. For a subjectivist wishing to make use of beta distributions, it is convenient that the theories of Jeffreys and Perks lead to a particular beta distribution since he will be in at least *some* measure of agreement with these two credibilists.

Let us consider a very simple example where subjective probability seems to be more appropriate than credibility. Suppose that we select a coin at random and there appears to be nothing wrong with it. We spin it and obtain r heads and s tails, and we must offer upper and lower odds for a head on the next spin, the smallest odds we would take and the largest we would give. Our Type II initial distribution is liable to be sharply peaked, with its mean very close to 0.5. In view of the previous discussions, I shall assume that it is sensible to take the Type II final probability as of the form $(r + k)/(r + s + 2k)$, and our problem is to select two values for k, say k_1 and k_2, which determine the upper and lower probabilities. The Laplace value $k = 1$ is certainly too small. We can decide on reasonable values for k_1 and k_2 by means of the "device of imaginary results" (Good,[27] pp. 35, 70). We imagine, for example, that an experiment has yielded exactly 1 tail and r heads, and ask ourselves, "How large must r be for which we would just give odds of 2:1 (probability 2/3) on a head next time?" Suppose that we decide, for one reason or another, by model-building, experience, discussion, and intuition, that the appropriate value of r would be between 25 and 40. Then we must take $k_1 = 23$, $k_2 = 38$. For this application, the question whether a credibilist should use $k = 1/2$ or $k = 1$ appears to be beside the point; the vague judgments of the subjectivist are much more pertinent.

Some further theory concerning simple sampling is implicit in Chapter 4, which deals with multiple sampling. One implication of those discussions is that the subjectivist might select a "Type III distribution" for k; in the last example it would be concentrated in the range $23 \leq k \leq 38$.

The device of imaginary results helps you to decide on an initial Type II distribution before sampling; this to some extent decreases the danger of unconscious cheating.

Addendum. In the course of correspondence with W. Perks, while this monograph was in press, he stated that his practical statistical philosophy is that of a subjectivist. His work on credibilism is no more than an attempt to give formal expression to ignorance, as a contribution to the logic of inference. In fairness to Jeffreys, it should be added that, for the problem of the coin discussed at the top of the page, he too would undoubtedly not use $k = 1/2$; he would I think be forced reluctantly into subjectivism.

4. Multiple Sampling

A *multiple sample* consists of a sequence of trials, each of which results in one of the "letters" of an "alphabet" consisting of a finite number t of letters. There is of course no real loss of generality if we denote the letters of the alphabet by $0, 1, 2, \cdots, t - 1$, or by $1, 2, \cdots, t$. The result of a (finite) sample is a finite sequence consisting of N letters, where N is called the *sample size*.

A *multinomial sample* is a multiple sample that results in a *random* sequence, that is, a sequence in which the successive trials are statistically independent. The probability that a multinomial sample of size N will have the *frequency count* (n_i) $(i = 1, 2, \cdots, t)$, is $N! \prod_i (p_i^{n_i}/n_i!)$ where p_i is the physical probability that any particular trial will result in the letter i. When the physical probabilities (p_i) are unknown, they might have a Type II initial probability distribution in the simplex $\sum p_i = 1$, from which the Type II probability of any particular frequency count (n_i) could be inferred by integration. The converse proposition is a generalization of de Finetti's theorem (Theorem 3.1), and we shall state and prove this converse. More general results are known (see, for example, Hewitt and Savage[56] and Freedman[25]), but the proofs of these results are much more complicated.

Let us say, following W. E. Johnson again, that a stochastic sequence, in a t-letter alphabet, is *permutable* if, for every possible frequency count (n_i), every sequence having this frequency count has the same probability, which we denote by $Q(n_1, \cdots, n_t)$. The generalized form of de Finetti's theorem, for multiple sampling, can be stated thus:

21

Theorem 4.1

A multiple sample that generates a permutable sequence can always be regarded as having been obtained by multinomial sampling in which the Type I probabilities (p_i) have a joint Type II initial distribution. Moreover this Type II distribution is unique if it is continuous. In short, permutable multiple sampling is multinomial sampling with unknown category probabilities.

The philosophical interpretation of this theorem is of course precisely analogous to that of the special case of simple sampling, discussed earlier.

Proof:

We clearly have the consistency condition, mentioned by W. E. Johnson,[67] that

$$Q(n_1 + 1, n_2, \cdots, n_t) + Q(n_1, n_2 + 1, \cdots, n_t) + \cdots$$
$$+ Q(n_1, n_2, \cdots, n_t + 1) = Q(n_1, n_2, \cdots, n_t) \quad (4.1)$$

Let $\Delta_1, \Delta_2, \cdots, \Delta_t$ be difference operators with respect to the t arguments, for example,

$$\Delta_1 Q(n_1, n_2, \cdots, n_t) = Q(n_1, \cdots, n_t) - Q(n_1 + 1, \cdots, n_t)$$

Then, by Equation 4.1, we have

$$\Delta_1 Q(n_1, \cdots, n_t) = Q(n_1, n_2 + 1, \cdots, n_t) + Q(n_1, n_2, n_3 + 1, \cdots, n_t)$$
$$+ \qquad Q(n_1, n_2, \cdots, n_t + 1) \geq 0$$

Therefore,

$$\Delta_1^2 Q(n_1, \cdots, n_t)$$
$$= \Delta_1 Q(n_1, n_2 + 1, \cdots, n_t) + \Delta_1 Q(n_1, n_2, n_3 + 1, \cdots, n_t) + \cdots$$
$$+ \Delta_1 Q(n_1, n_2, \cdots, n_t + 1)$$
$$= \text{sum of } (t - 1)^2 \text{ terms, each a } Q$$

Similarly, for all q ($q = 0, 1, 2, 3, \cdots$),

$$\Delta_1^q Q(n_1, \cdots, n_t) = \text{sum of } (t - 1)^q \text{ terms, each a } Q$$

and more generally

$$\Delta_1^{q_1} \Delta_2^{q_2} \cdots \Delta_{t-1}^{q_{t-1}} Q(n_1, n_2, \cdots, n_t)$$
$$= \text{sum of } (t - 1)^{q_1 + \cdots + q_{t-1}} \text{ terms, each a } Q \geq 0$$

Therefore, the multiple sequence $\{Q(n_1, n_2, \cdots, n_{t-1}, 0)\}$ is "totally monotonic" in the generalized sense of Hildebrandt and

Schoenberg,[57] and so, by Theorem 1 of that paper, $Q(n_1, \cdots, n_{t-1}, 0)$ is of the form

$$\int_0^1 \int_0^1 \cdots \int_0^1 p_1^{n_1} \cdots p_{t-1}^{n_{t-1}} \, d_1 \, d_2 \cdots d_{t-1} F(p_1, \cdots, p_{t-1})$$

where F is monotonic increasing; that is, $F(p_1, \cdots, p_{t-1}) \geq F(p_1', \cdots, p_{t-1}')$ whenever $p_1 \geq p_1', \cdots, p_{t-1} \geq p_{t-1}'$. Also F is unique in the sense that any two F's can differ only in an enumerable number of hyperplanes. If F is continuous, then it is strictly unique. By putting all the n_i's equal to zero, we now see that F is a distribution function. With the help of Equation 4.1, we can now prove inductively that

$$Q(n_1, \cdots, n_t) = \int_0^1 \int_0^1 \cdots \int_0^1 p_1^{n_1} \cdots p_{t-1}^{n_{t-1}} (1 - p_1 - \cdots - p_{t-1})^{n_t}$$
$$\times d_1 \, d_2 \cdots d_{t-1} F(p_1, \cdots, p_{t-1})$$

and the theorem is established.

4.1 Multinomial estimation

The extension (Theorem 4.1) of de Finetti's theorem provides, for the extreme subjectivist, a definition of the physical probabilities of a t-category classification. This clears the ground for the discussion of multinomial estimation, that is to say, the problem of estimating these probabilities. The theorem makes it uncontroversial that the problem exists, but there is bound to be controversy about how the estimation is to be done.

As in the rest of this monograph, we shall be concerned primarily with effectively small samples, samples for which at least one of the sample frequencies n_i is small, including values as small as $n_i = 1$ and $n_i = 0$. For effectively large samples, what we shall say is still relevant but of less interest.

For a sample of size N, the maximum-likelihood estimate of p_i is n_i/N. When n_i is small compared with N/t, this estimate will usually be too small for betting purposes, especially if $n_i = 0$. It will also be too small as an estimate of p_i when the purpose of making the estimates is for use in deciding whether another sample comes from the same population. In Chapter 5, we say more concerning this decision problem and concerning the closely allied problem of discrimination.

Many statisticians faced with this estimation problem would make some *ad hoc* modification to the observed frequencies. That is, they would replace n_i by some slightly different number n_i^*, and use n_i^*/N as their estimate of p_i, provided of course that they wished to make a point estimate. It is natural to move all the n_i's closer to the mean N/t, or, as one might say, to "squash" the frequency count. This seems a natural thing to do without explicit reference to intuitive probability. For the expected scatter of the numbers n_i/N, as measured, for example, by the Type I expectation of

$$\sum \left(\frac{n_i}{N} - \frac{1}{t} \right)^2$$

exceeds that of the p_i's. Among the various possible "squashing" methods, linear squashing seems the simplest, in which the n_i^*'s are selected so that

$$n_i^* - \frac{N}{t} = \lambda \left(n_i - \frac{N}{t} \right)$$

where λ is some positive constant less than unity. If, owing to some lack of symmetry in the problem, the initial Type II probabilities were q_i, then the corresponding squashing rule would be

$$n_i^* - Nq_i = \lambda(n_i - Nq_i)$$

but, for the sake of simplicity, we shall restrict our attention to the case $q_i = 1/t \ (i = 1, 2, \cdots, t)$. If we took $\lambda = 1$, we should be taking the sample "at its face value" and using the maximum-likelihood estimates of the physical probabilities. We shall consider methods later for the selection of λ when linear squashing is used.

A certain amount of rational justification for linear squashing was apparently first given by W. E. Johnson.[67] He used the probability estimate

$$\frac{n_i + k}{N + kt} \tag{4.2}$$

so that k can be regarded as a *flattening constant* which is added to each of the sample frequencies. Flattening and squashing clearly come to the same thing with

$$\lambda = \frac{N}{N + tk} \tag{4.3}$$

Before discussing Johnson's argument, let us go back a few years to the work of the actuary Lidstone.[80] He suggested that the natural generalization of Bayes' postulate would be a uniform initial Type II distribution of the physical probabilities in the simplex $\sum p_i = 1$, and found, by means of Dirichlet's multiple integral (Whittaker and Watson,[110]

p. 258) that this implied a flattening constant $k = 1$. In other words, the natural generalization of Laplace's law of succession is the set of estimates

$$p_i = \frac{n_i + 1}{N + t} \qquad (4.4)$$

(We shall throughout use the symbol p_i or $\hat{p}_i$ for an estimate of p_i, not necessarily the same estimate on each occasion.) By virtue of the extension of de Finetti's theorem (Theorem 4.1), the estimates, Equation 4.2, also imply the Bayes postulate of a uniform initial Type II distribution in the simplex, provided that this distribution is assumed to be continuous. More generally, the use of the flattening constant k is equivalent to that of the Type II initial distribution of the symmetrical Dirichlet form with probability element

$$\frac{\Gamma(tk)}{\Gamma(k)^t} \prod p_i^{k-1} \, dp_1 \, dp_2 \cdots dp_{t-1} \qquad (4.5)$$

in the region $p_1 + p_2 + \cdots + p_{t-1} < 1 \, (p_t = 1 - p_1 - \cdots - p_{t-1})$. The terminology "Dirichlet distribution" is that used by Wilks;[114] when the distribution is not necessarily symmetrical, the density is of the form

$$\Gamma(\textstyle\sum k_i) \prod_i \left\{ \frac{p_i^{k_i-1}}{\Gamma(k_i)} \right\} \qquad (4.6)$$

For the case $t = 2$, it reduces of course to the beta distribution. It has frequently occurred in statistical literature: for example, Lidstone,[80] Jeffreys,[63] Perks,[93] Good,[29, 33] Mauldon,[83] N. L. Johnson,[65] Mosimann,[87, 88] and Wilks.[114] The use of this general Dirichlet distribution is equivalent to that of a set of flattening constants (k_i). Even greater flexibility can be conveniently achieved by using a linear combination of such distributions. Some such linear combinations will of course be again symmetrical. In this manner we could, for example, construct convenient symmetrical Type II distributions having the property that one of the p_i's would be likely to be much larger than the others, which one being unspecified in advance.

W. E. Johnson seems to have been unaware of the connection between the use of a flattening constant k and the symmetrical Dirichlet distribution. He did however make two distinct contributions to the theory of multiple sampling. In Johnson,[66] p. 183, he proposed what he called the "Combination Postulate," that "each [ordered] partition $(n_1, n_2, \cdots, n_t)$ of N is equiprobable." He should have made some such qualification as "provided that the null hypothesis is not (approximately) true," where by the "null hypothesis" we mean for the moment that the p_i's are all equal. For he says (Johnson,[66] pp. 188–189), in relation to an example, "the die is not known to be either physically or geometrically regular." This led him to an additive constant $k = 1$. But later (Johnson[67]), he produced some new theory which he regarded as superior. He abandoned the combination postulate, and instead he made the permutation postulate together with the following one, which I shall call *W. E. Johnson's sufficiency postulate:*

> *The credibility that the next trial will be of category i depends only on n_i and N ($i = 1, 2, \cdots, t$).*

From his permutation and sufficiency postulates, he deduced that this credibility (the Type II expectation of p_i, in our terminology) is given by flattening, with some flattening constant k, as in Equation 4.2, but he gave no suggestions for determining the constant. The kernel of his proof was to deduce from Equation 4.1 that, if the credibility be denoted by $f(n_i, N)$, we have

$$f(n_i + 1, N) - f(n_i, N) = f(n_j, N) - f(n_j - 1, N) \qquad (4.7)$$

for all i and j where $i \neq j$. From this he deduced that $f(n_i, N)$ is linear in n_i, and the final conclusion then followed without much difficulty. It seems not to have been previously noticed that this argument breaks down when $t = 2$, since the deduction of linearity from Equation 4.7 can no longer be made. That there is something wrong when $t = 2$ can also be seen in the following manner.

We know that the use of a flattening constant is equivalent to that of an initial Type II density of the symmetrical Dirichlet form proportional to $\prod p_i^{k-1}$, provided that the distribution is continuous. Therefore, when $t = 2$, Johnson's sufficiency postulate implies that the Type II initial probability distribution of p_1 is a beta distribution if his argument is correct. But when $t = 2$, *Johnson's sufficiency postulate is tautologically true.* Since the initial Type II distribution could have been any arbitrary *physical* distribution, for an appropriate Type II sampling scheme, without violation of the permutation postulate, it is clear that Johnson's argument *proves too much* when $t = 2$. This is most unfortunate, since, if G. F. Hardy's use of the beta distribution could be made uncontroversial, Hume would be defeated and statistics could begin.

At least one argument can be adduced against Johnson's sufficiency postulate. In the sampling of species, discussed in Chapter 8, there is a different, and uncontroversial, method of estimating the physical probabilities that is inconsistent with the use of a flattening constant. In this problem the probability estimates p_i are actually expressed in terms of the *frequencies of the frequencies* of the categories, and this would not be allowable if the sufficiency postulate were valid. It is a necessary condition for the validity of the theory of the species-sampling problem that the number of categories having frequencies 1 and 2 should be large, and consequently that the sample size is very large (though still *effectively* small). Johnson's sufficiency postulate does not seem unreasonable when these conditions are not met. (See also Appendix E.)

One argument for the Type II initial distribution of the symmetrical Dirichlet form is that it is more general and flexible than Bayes' postulate (which is the special case $k = 1$) and so gives reasonable scope for the statistician to take his initial judgments into account, although it also gives him scope for cheating. One method of taking these judgments into account is to make a guess of the initial expectation, $\hat{\rho}_0 = \mathscr{E}_{\mathrm{II}}(\rho)$, of the "repeat rate"

$$\rho = p_1^2 + p_2^2 + \cdots + p_t^2$$

and to equate this expectation to $(k + 1)/(kt + 1)$. (See Chapter 5.) This gives

$$k = \frac{1 - \hat{\rho}_0}{\hat{\rho}_0 t - 1} \tag{4.8}$$

This is a special case of the following general procedure: *The Bayesian could select one of an m-parameter class of initial distributions by guessing initial values of m population parameters.* If he guesses more population parameters than he has in his class of initial distributions, then he might be forced to modify his guesses or to enlarge the class. This is one deduction from the thesis that the Bayesian should make use of *mature* judgments. (Good,[27] p. 4.)

Instead of guessing a value for $\hat{\rho}_0$, you could guess upper and lower bounds for it, or preferably a Type III initial distribution for it. This would imply a Type III distribution, say $D(k)$, for k, and this would convert the *composite* Type II (Dirichlet) hypothesis, the disjunction of the simple Type II Dirichlet hypotheses, into a *simple* statistical hypothesis of Type II, having the density

$$\frac{1}{(t - 1)!} \int_0^\infty \Gamma(tk)\Gamma(k)^{-t} \prod p_i^{k-1} \, dD(k) \tag{4.9}$$

In physical terms, if we do a Type III sampling of k with distribution $D(k)$, followed by a Type II sampling of (p_i) with the symmetrical Dirichlet distribution of parameter k, followed by a Type I sampling of (n_i), all this I say comes to the same as a Type II sampling of (p_i) with the density function, Equation 4.9, followed by the sampling of (n_i).

Likewise, if the distribution of k is subjective, and the composite Dirichlet sampling is credibilistic, then the effect is the same as if a subjective simple Type II distribution had been assumed. This is not a farfetched possibility, since W. E. Johnson's sufficiency postulate might seem reasonable to many people, as an approximation to be used with care, whereas a distribution for k will, I think, seem much more subjective. There is a hierarchy: psychological → subjective → logical → physical, but the boundaries between the four elements of the hierarchy are not clear-cut.

Compromises between Bayesian and non-Bayesian methods were discussed in Chapter 2 and will be discussed again in Chapter 5. One such compromise is to estimate k by Type II maximum likelihood on the basis of the sample (n_i). A partial defense for such a procedure is that the main contribution to an integral as in Expression 4.9 is liable to be made by values of k in the neighborhood of the value of k that maximizes the integrand. A similar partial justification exists of course for ordinary maximum likelihood: it is roughly Bayesian.

The ordinary Type I maximum-likelihood estimate of p_i corresponds to $k = 0$ and therefore to the divergent symmetrical Dirichlet initial Type II density proportional to

$$\prod p_i^{-1} \tag{4.10}$$

provided of course that the maximum-likelihood estimates are interpreted as betting probabilities. Then, by Equation 4.8, the initial expectation of ρ must be 1. In other words, one of the p_i's must be 1 and the rest 0. This merely verifies that maximum-likelihood estimates can hardly be interpreted as betting probabilities.

A form of inference, similar to fiducial inference, and to which similar criticisms apply, was suggested for the multinomial distribution by A. D. Roy.[98] (See p. 11 of this monograph.) He called it "pistimetric inference." Although he does not say so explicitly, a perusal of his paper shows at once that he is *implicitly* making use of the initial Type II distribution just defined in Equation 4.10. The pistimetrician behaves like a Bayesian who uses an unreasonable initial distribution. It is better to select one or two not unreasonable, somewhat arbitrary distributions. When $t = 2$, Equation 4.10 reduces to Haldane's distribution, which cannot be the best of the symmetrical beta distributions any more than Equation 4.10 can be the best of the Dirichlets.

It will be recalled that the credibilists, Jeffreys and Perks, suggested the flattening constant $k = 1/2$ for the case $t = 2$. They have generalized their suggestions to arbitrary values of t, for permutable multiple sampling when the initial information concerning the t categories is mere symmetry. Jeffreys,[63] using his invariance theory, arrived at the value $k = 1/2$ for all values of t, whereas Perks[93] advocated $k = 1/t$. Perks' invariance theory applies only when the Type I population depends on a single parameter p and so is inapplicable when $t > 2$. But he argued in effect as follows.

Suppose that t is even, $t = 2u$. We could suppose that the "letters" $1, 2, \cdots, u$ are lumped together, and that the letters $u + 1, \cdots, 2u$ are also lumped together. Then we would have a two-category classification for which the flattening constant would be 1/2. (The initial information concerning the two categories is symmetrical.) If there is a flattening constant for the original t-category classification, then this constant must be $1/t$, for the sake of consistency of the final betting probabilities. (But see Section 4.2.)

In his contribution to the discussion of Perks' paper, Jeffreys says, in relation to his own invariance theory, that his theory is not yet in optimal form when there are several parameters. In Jeffreys,[63] p. 184, he mentions his flattening constant $k = 1/2$, but in a somewhat incidental manner, and without reference to Perks' suggestion.

The following argument seems, at first sight, to lend further support for Perks' estimate as compared with Jeffreys'. (This argument is slightly implicit in Perks' paper.) Let us imagine that in N trials where N is large, we have $n_1 = N - 1$. For what value of M would we just be prepared to have an even bet that the next M letters will all be 1's? I think that most readers would agree that, if betting be compulsory and if further background of any particular application is to be ignored, then M should lie somewhere between $N/4$ and $3N/4$. But, if there is a flattening constant k, the final Type II probability that the bet will be won is (see Equation 4.2)

$$\frac{N - 1 + k}{N + tk} \quad \frac{N + k}{N + 1 + tk} \quad \frac{N + M - 2 + k}{N + M - 1 + tk}$$

$$= \frac{\Gamma(N + M - 1 + k)\Gamma(N + tk)}{\Gamma(N + M + tk)\Gamma(N - 1 + k)} \sim \left(\frac{N}{N + M}\right)^{k(t-1)+1} \quad (4.11)$$

If this is equal to 1/2, then

$$0.24 < (t - 1)k < 2.1 \quad (4.12)$$

On the basis of this argument, a credibilist should apparently prefer Perks' flattening constant to Jeffreys'. But we shall see later, for example, on pages 33, 35 to 41, and 44 to 46, that different frequency counts $(n_1, n_2, \cdots, n_t)$ make different flattening constants seem reasonable. Formula 5.9, when applied to the present example, gives a flattening constant that tends to Perks' value, $1/t$, when N tends to infinity.

4.2 Natural Groupings of the Categories

We conclude this chapter with a warning that the use of a symmetrical Dirichlet distribution leads to inconsistencies when there is any natural grouping together of the categories. The argument is based on the consideration of arboresque (treelike) classifications. Consider, for example, a dichotomous tree for which the physical probabilities in the first generation are p_1 and p_2 ($p_1 + p_2 = 1$), those in the second generation are p_{11}, p_{12}, p_{21}, and p_{22} ($p_{11} + p_{12} = p_1, p_{21} + p_{22} = p_2$), etc. Let the Type II initial probability density in the νth generation be

$$\text{P.D.}(p_{i_1 i_2 \cdots i_\nu} = x) = f_\nu(x)$$

Assume that $p_{i_1 i_2 \cdots i_{\nu+1}}$ has the conditional initial density

$$\text{P.D.}(p_{i_1 i_2 \cdots i_{\nu+1}} = x \mid p_{i_1 i_2 \cdots i_\nu}) = f(x)$$

this density function being mathematically independent of ν. Then one can see that

$$f_{\nu+1}(x) = \int_x^1 \frac{1}{y} f_\nu(y) f\left(\frac{x}{y}\right) dy \tag{4.13}$$

This is a "Mellin convolution," and it follows that

$$f^*_{\nu+1}(u) = f^*_\nu(u) f^*(u) = f^*(u)^{\nu+1} \tag{4.14}$$

where the asterisk denotes a Mellin transformation,

$$f^*(u) = \int_0^\infty x^{u-1} f(x) \, dx \tag{4.15}$$

where the upper limit of integration can here be replaced by 1. If then $f(x)$ is a symmetrical beta function with parameter k, we have

$$f^*_\nu(u) = \left\{ \frac{\Gamma(2k)\Gamma(k + u - 1)}{\Gamma(k)\Gamma(2k + u - 1)} \right\}^\nu \tag{4.16}$$

so that $f_\nu(x)$ is a Meijer G-function (see, for example, Erdélyi and others,[20] p. 374). If $k = 1$, that is, if $f(x) = 1$, we have

$$f_\nu(x) = \frac{1}{(\nu - 1)!} (- \log x)^{\nu - 1} \tag{4.17}$$

and whatever the value of k, the density function $f_\nu(x)$ is not a beta function for $\nu > 1$, as it would have to be if the joint distribution of the νth-generation physical probabilities were of the Dirichlet form.

It is interesting to note that a 2×2 contingency table gives rise to two distinct arboresque classifications. The question arises whether conditions of consistency on the Type II initial distributions can shed light on what these distributions must be. This question is investigated in Appendix E, where it is found that the indiscriminate use of Johnson's sufficiency postulate leads to a contradiction.

5. Multinomial Discrimination and Significance

5.1 Discrimination Between Two Simple Statistical Hypotheses

If H_1 and H_2 are two simple statistical hypotheses and if E is an observational or experimental result whose probabilities, given H_1 and given H_2, are specified by the definitions of H_1 and H_2, then the simple likelihood ratio

$$\frac{P(E \mid H_1)}{P(E \mid H_2)} \tag{5.1}$$

is equal to the factor by which the initial odds of H_1 must be multiplied in order to get the final odds,

$$\frac{O(H_1 \mid E)}{O(H_1)} \tag{5.2}$$

The definition of odds is $p/(1-p)$, where p is probability. We have taken the proposition "H_1 or H_2" for granted, and omitted it from the notation, but strictly it should appear to the right of all the vertical strokes.

Turing called this factor the "factor in favor of H_1 as against H_2 provided by E," which I denote by $F(H_1/H_2:E)$. (See Good,[27] Ch. 6.) The factor is also used by Jeffreys,[63] but its use is slightly obscured since he consistently takes the initial odds as 1, so that the factor becomes the final odds. It is here called the "Bayes factor." The oblique stroke can be read "as against" and should not be confused with the vertical stroke meaning "given." The colon can be read "provided by" and it too is not equivalent to a vertical stroke. These three symbols have these meanings only when they separate expressions that denote propositions.

The factor is defined in this manner even if H_1 and H_2 are composite, in which case the non-Bayesian will not use the symbols $P(E \mid H_1)$ and $P(E \mid H_2)$. It is only when both hypotheses are simple statistical hypotheses that the Bayes factor is equal to the likelihood ratio. Otherwise the likelihood ratio is defined as a ratio of *maximum* likelihoods. But even in the case of simple statistical hypotheses the terminology "factor in favor of a hypothesis" has the advantage of emphasizing the Bayesian interpretation and has immediate intuitive appeal.

When one wishes to discriminate between two simple statistical hypotheses, the use of the likelihood ratio is uncontroversial, owing to the Neyman-Pearson lemma (for example, see Wilks[114]). The Bayesian regards this statistic as the best, because it is the factor in favor of a hypothesis. (We are here ignoring the possibility that in some circumstances there might be other statistics that are a little faster to compute, without losing much power.)

We now turn our attention to multinomial distributions. Let H_p be the hypothesis that a t-category multinomial distribution has Type I probabilities $(p_1, p_2, \cdots, p_t)$, and let H_q be similarly defined. We wish to discriminate between these two hypotheses by means of a sample E with frequency count (n_i). It is clear that the logarithm of the factor is

$$W(H_p/H_q : E) = \sum n_i \log \left(\frac{p_i}{q_i} \right) \tag{5.3}$$

where I use the symbol W for "weight of evidence," which is synonymous with "log-factor." (The expected log-factor, given either hypothesis, is of course expressible in terms of entropy and crossentropy.)

If one or both of the two populations (p_i) and (q_i) are known only by means of finite basic samples, then this method of obtaining a log-factor can be misleading. This can happen, for example, if the maximum-likelihood estimates of the p_i's and q_i's are used. More specifically, if the basic samples are (r_i) and (s_i), where $\sum r_i = R$, and $\sum s_i = S$, and we wish to discriminate whether a third (nonbasic) sample (n_i) is from the first population (hypothesis H_p) or from the second one, then the use of the maximum-likelihood estimates would lead to the apparent log-factor

$$y = \sum n_i \log \left(\frac{r_i/R}{s_i/S} \right) \tag{5.4}$$

in favor of H_p as against H_q. That this can be misleading is shown by the following example: Suppose that the sample (s_i) is infinite, that $q_i = 1/t$ for all i, and that the p_i's are close to $1/t$. Let

$$\chi^2 = \sum \frac{(r_1 - R/t)^2}{R/t}$$

Then it can be shown that the ordinary (Type I) expectation of the apparent log-factor y when H_p is true, is given approximately by

$$\mathcal{E}(y \mid H_p) \approx \frac{N}{2R} \{\mathcal{E}(\chi^2) - 2(t - 1)\} \tag{5.5}$$

whereas the expectation of the true (but unknown) log-factor is given approximately by

$$\mathcal{E}\{W(H_p/H_q : E) \mid H_p\} \approx \frac{N}{2R} \{\mathcal{E}(\chi^2) - (t - 1)\} \tag{5.6}$$

If $\mathcal{E}(\chi^2) < 2(t - 1)$, the expected apparent log-factor is actually negative. (The proofs of these statements, and of the remaining ones in this paragraph, will be omitted. They were derived jointly by J. W. S. Cassells, G. C. Wall, and the writer, during the years 1940–1948 in unpublished work.) Matters can be improved by squashing or flattening the basic samples. If no flattening is used, the ratio of the expected apparent log-factor to the expected true log-factor is

$$1 - \frac{t - 1}{\mathcal{E}(\chi^2) - (t - 1)} \tag{5.7}$$

whereas the ratio is

$$1 - \frac{t - 1}{\mathcal{E}(\chi^2)} \tag{5.8}$$

if the flattening constant

$$\frac{R}{t} \cdot \frac{t - 1}{\mathcal{E}(\chi^2) - (t - 1)} \tag{5.9}$$

is applied to the basic sample (r_i). Moreover, any other flattening constant produces a smaller ratio than this one produces. In order to compute this flattening constant, it is necessary to estimate $\mathcal{E}(\chi^2)$ by the observed value of χ^2, or in some other manner. Note too that the idea of maximizing the expected *apparent* log-factor has not been justified by showing that it leads to optimal discrimination, even approximately. Further research on this question would be of interest.

The squashing constant corresponding to the flattening constant, Equation 5.9, is

$$\lambda = 1 - \frac{t-1}{\mathscr{E}\chi^2} \qquad (5.10)$$

the ratio of the expected chi-squared "bulge" to $\mathscr{E}(\chi^2)$. It is intuitively reasonable to suppose that, if the second basic sample (s_i) is not infinite, then the same flattening rule could be applied to it, using its own value of χ^2.

The formulas for the flattening and squashing constants (Equations 5.9 and 5.10) make sense only if $\chi^2 > t - 1$. If $\chi^2 < t - 1$, one would, for consistency, use a flattening constant, $k = \infty$, in other words adopt the hypothesis of flat randomness.

We note for future reference that, when $R \to \infty$, the flattening constant, Expression 5.9, almost certainly tends to

$$\frac{t-1}{t(t \sum p_i^2 - 1)} \qquad (5.11)$$

If a flattening constant is used, it might seem at first sight that the log-factor in favor of H_p would be

$$\sum n_i \log \frac{(r_i + k)/(R + tk)}{(s_i + k')/(S + tk')} \qquad (5.12)$$

where k and k' are the flattening constants for the two basic samples. But it would be more accurate to think of the nonbasic sample's arriving one "letter" at a time, with adjustment of the flattening constants after the arrival of each "letter." On hypothesis H_p the nonbasic sample would be used to increase the first basic sample, and on H_q, to increase the second basic sample.

We shall later return to this problem of discrimination from another point of view, but first it is necessary to clear the ground by discussing some philosophy.

5.2   Bayes/Non-Bayes Compromises †

The extreme subjectivist believes that by enough self-interrogation, with a free use of the device of imaginary results, and honest mature detached self-critical judgments, he could write down a numerical probability $P(A \mid B)$ for any propositions A and B. The trouble is that this would take too long, and the possibility of doing it is gravely hindered by the vagueness of language (see Good,[27] p. 48). Conse-

† See also Section 2.2.

quently a compromise with non-Bayesian methods is often forced upon the Bayesian. The following example of such a compromise is described in Good,[33] p. 863. First a Bayesian model can be used for the selection of a statistic in the form of a Bayes factor F. But the Bayesian might not be sure of his model, so he decides to use F as a statistic in the non-Bayesian manner, and he calculates a tail-area probability P, given the null hypothesis. The Bayesian might be surprised if F does not lie in the range $(1/30P, 3/10P)$; if so, he would presumably carry out some more self-interrogation. He can do this just as well with an imaginary experimental result as with a real one.

The Bayesian who is prepared to believe that a credibility distribution exists would take this as his own subjective probability distribution if he knew what it was. It is then natural for him to select Type II hypotheses that are plausible credibility distributions and to assume a Type III probability distribution over the space of these Type II hypotheses. Moreover, he might judge that the integration to which this would lead could be adequately approximated by using Type II maximum likelihood, that is, to select the Type II hypothesis that gives to the observations their largest probability. The use of Type II maximum likelihood is another example of a compromise between (modern) Bayesian and non-Bayesian methods.

Again a Bayesian might reject a Type II null hypothesis in the light of enough experimental evidence. In other words, he could use Type II significance tests as well as Type II estimation of parameters. There is also the possibility of a Type II likelihood-ratio test.

If the Bayesian prefers, he can think of a credibility distribution as a hypothetical physical distribution, since he can, with some boggle, imagine an infinite sequence of distinct universes selected at random and define a credibility as an (almost certain) limit of proportional frequencies in these universes. The notion of a random selection of universes is of course purely metaphysical, but not, I think, self-contradictory; and any crutch to one's judgment can be used unofficially. It might be inexpedient to mention to one's customers that one had such naughty unscientific private thoughts.

Most of the remaining methods of this chapter involve a compromise of one kind or another.

5.3 Estimation of a Flattening Constant and Tests of Equiprobability

In Chapter 4 we were interested in the estimation of the Type I probabilities (p_i) in terms of a sample of frequency count (n_i), where $\sum n_i = N$, and one of the methods of estimation was to assume a flattening constant k. We pointed out that in some circumstances the flattening constant cannot be used, but no general rule was given determining these circumstances. We also had difficulty in deciding what value of k should be used to represent ignorance. If the Type II distribution of (p_i) is assumed to be physical, then the question of whether k exists is one of significance testing, and the selection of k is a problem of estimation. It seems to me that the same is true even if the Type II distribution is intuitive.

W. E. Johnson's sufficiency postulate is not convincing enough to give complete assurance that ignorance can be represented by a flattening constant, that is, by a symmetrical Dirichlet distribution. And Jeffreys' and Perks' invariance arguments are, for $t > 2$, mutually contradictory and not convincing enough to determine the values $k = 1/2$ or $1/t$. But Johnson's postulate can reasonably be used as a justification for setting up the composite Type II null hypothesis that the Type II distribution of (p_i) is a symmetrical Dirichlet distribution, and the arguments of Jeffreys and Perks lead to the simple Type II null hypotheses that this distribution has $k = 1/2$ or $k = 1/t$. The values $k = 0$ and $k = 1$ might also be given special consideration, although the arguments for them seem less cogent. They are Type II null hypotheses of smaller (subjective) Type III initial probability. Finally, the value $k = \infty$ implies merely that all p_i's equal $1/t$, and this is a familiar *Type I* null hypothesis. These then are the five credibilist's flattening constants, $k = 0, 1/t, 1/2, 1,$ and ∞.

Let us first assume that the Type II distribution of (p_i) is a symmetrical Dirichlet density, proportional to $\prod p_i^{k-1}$. The Type II likelihood of any value of k can be readily derived from Dirichlet's integral,

$$\int \cdots \int_{\Sigma p_i = 1} \prod_{i=1}^{t} p_i^{k_i - 1} \, dp_1 \cdots dp_{t-1} = \frac{\prod \Gamma(k_i)}{\Gamma(\sum k_i)} \tag{5.13}$$

or otherwise (see Equation 3.6). The (Type II) likelihood is found to be

$$P((n_i) \mid k) = \frac{N!}{\prod n_i!} \frac{\prod (n_i - 1 + k)^{(n_i)}}{(N - 1 + tk)^{(N)}}$$

$$= \frac{\Gamma(tk)\Gamma(N + 1) \prod \Gamma(n_i + k)}{\Gamma(k)^t \Gamma(N + tk) \prod \Gamma(n_i + 1)} \tag{5.14}$$

where $x^{(n)} = x(x - 1) \cdots (x - n + 1)$. (Note the check that, when $N = 0$, the probability is 1.) We can immediately reject the hypothesis that $k = 0$, which is *equivalent* to the Type I maximum-likelihood estimation of the p_i's, since, unless all but one of the n_i's vanish, we have $P((n_i) \mid k = 0) = 0$ if $N > 0$. Let us take a numerical example, and plot the Type II likelihood of the sample. Take $t = 5, n_1 = 1, n_2 = 2, n_3 = 3, n_4 = 4, n_5 = 5$, so that $N = 15$. We find that $L(k) = -\log_{10} P((n_i) \mid k)$ takes the following numerical values: $L(0) = \infty$, $L(1) = 3.59$, $L(2) = 3.26$, $L(3) = 3.14$, $L(7) = 2.99$, $L(\infty) = 2.91$. This last value was calculated from the formula

$$P((n_i) \mid k = \infty) = \frac{N! t^{-N}}{\prod n_i!} \qquad (5.15)$$

It seems that, in this example, the symmetrical Dirichlet Type II hypothesis of maximum Type II likelihood occurs when $k = \infty$. (This value of k is equivalent to the Type I null hypothesis $p_i = 1/t$ ($i = 1, 2, \cdots, t$) as we said before.) This is a special case of the following conjecture:

Conjecture. Given a Type I sample (n_i), the Type II log-likelihood of the flattening constant k ($k > 0$), that is, of the symmetrical Dirichlet Type II distribution of parameter k when regarded as a function of k, has at most one local maximum. It takes its maximum at $k = \infty$, if $\chi^2 < t - 1$, and for a finite value of k if $\chi^2 > t - 1$, where

$$\chi^2 = \frac{t}{N} \sum \left(n_i - \frac{N}{t} \right)^2 \qquad (5.16)$$

Some support for a part of this conjecture is provided by the following argument. We have

$$\frac{d}{dk} \log_e P((n_i) \mid k) = t g(tk) + \sum g(n_i + k) - t g(N + tk) - t g(k) \qquad (5.17)$$

where

$$g(z) = \frac{\Gamma'(z)}{\Gamma(z)} \sim \log_e z - \frac{1}{2z} + \sum_{r=1}^{\infty} \frac{(-1)^r B_r}{2 r z^{2r}} \quad \left(B_r = \frac{1}{6}, \frac{1}{30}, \frac{1}{42}, \cdots \right)$$

when $z \to \infty$. By expanding Equation 5.17 as far as the terms in k^{-2}, we see that, when k is large,

$$\frac{d}{dk} \log_e P((n_i) \mid k) \sim \frac{N}{2k^2 t} (t - 1 - \chi^2) \qquad (5.18)$$

Since the Type II log-likelihood is minus infinity when $k = 0$, the only unproved part of the conjecture is that there is at most one local maximum. I have found this to be true in five other numerical examples (with the help of an electronic computer).

In the numerical example just considered (with $N = 15$), we have $\chi^2 = 10/3$, and there are four degrees of freedom ($t - 1 = 4$). In the following example, $\chi^2 = 21.5$, again with four degrees of freedom. We take $t = 5$, $n_1 = 0$, $n_2 = 2$, $n_3 = 3$, $n_4 = 3$, $n_5 = 12$, $N = 20$. Some of the values of $L(k)$ are given in Table 5.1. In this example, the Type II

TABLE 5.1

k	$L(k)$	k	$L(k)$
0	∞	1.4	4.084
0.1	5.138	1.6	4.123
0.2	4.560	1.8	4.164
0.3	4.307	2.0	4.205
0.4	4.172	2.5	4.309
0.5	4.096	3.0	4.406
0.6	4.053	3.5	4.496
0.7	4.030	4.0	4.578
0.8	4.021	6.0	4.841
0.9	4.020	8.0	5.029
1.0	4.026	10.0	5.170
1.2	4.050	∞	6.131

maximum-likelihood value of k is about $k = 0.9$. It is interesting to note that this value of k is close to the value suggested in Equation 5.9 for different reasons. (I have found that this approximation is quite good in several other numerical examples, but I do not have a general proof.) All values of k between 0.5 and 1.6 have over one hundred times the likelihood of $k = \infty$. Provided that an appreciable amount of Type III probability is associated with the interval (0.5, 1.5), and the initial Type II probability that $k = \infty$ is not more than 1/2, then we would reasonably reject this hypothesis. For example, we might assume, given that k is finite, that the Type III initial distribution of $\log k$ is symmetrical about the origin, and that, for each positive number $\alpha < 1$,

$$P\left(\alpha < k < \frac{1}{\alpha}\right) = 1 - \alpha \qquad (5.19)$$

(The Jeffreys-Haldane divergent distribution, for which the density of k is proportional to $1/k$, will not do for this problem since it leads to *final* probability 0 for any finite interval of values of k.) Then the initial Type III probability density of k is

$$\phi(k) = \frac{1}{2} \qquad (0 < k < 1) \left.\begin{array}{c} \\ \\ \end{array}\right\}$$
$$\qquad = \frac{1}{2k^2} \qquad (1 < k) \qquad\qquad\qquad (5.20)$$

We find that the final Type III probability that k lies between 0.6 and 1.2 exceeds 0.5 and the Bayes factor against $k = \infty$ is somewhat greater than 100. One should of course not dogmatically assume this Type III distribution of k and should be prepared to allow for prior information in each application.

Let us compare this with some non-Bayesian methods. The tail-area probability for the ordinary χ^2 is about 1/5000 after allowing for the continuity correction (Cochran[11]), which reduces χ^2 to 20.25. The tabular chi-squared distribution is unreliable so far into the tail, and this applies also to the likelihood-ratio statistic

$$\mu = 2 \sum n_i \log_e n_i + 2N \log_e t - 2N \log_e N \qquad (5.21)$$

(See Wilks[113] and, for tables of $2n \log_e n$, see Ku[3][76].) Here $\mu = 20.144$, and the corresponding tail-area probability is 1/2800. It seems in this example that the likelihood-ratio statistic is better than χ^2, since the probability that one of the five cells will have an entry of 12 or more is 1/2300, and this clearly exhausts nearly all the evidence provided by the sample against the null hypothesis $k = \infty$. (The maximum-entry significance test is discussed, with further references, in Good.[33] It, of course, does not always exhaust nearly all of the information.)

An objective test of the hypothesis $k = \infty$, which does not depend on any asymptotic expansions, can be derived from the "empty cells" test (see, for example, Good,[33] p. 864). First write down the probability that there will be as many or more zero n_i's as the number observed. Then remove these empty cells, subtract 1 from all the other cells, and repeat the test with the reduced values of N, t, and n_i's. Continue this process until there is only one cell left. In this way we obtain a sequence of tail-area probabilities that need to be combined. In order to apply say Fisher's method of combining a series of tests (Fisher,[22] §21.1), strictly speaking one should make the distributions continuous. This can be done by adding continuously distributed random numbers between 0 and 1 to the numbers of empty cells at each stage of the iterated empty-cell test. (This method of making discrete distributions

continuous is used, for example, in E. S. Pearson[91] and Kincaid.[73])
If the number of empty cells is s, denote the tail-area probability, for
assigned N and t, by $P(s \mid N, t)$. In the preceding example, the tails are
$P(1 \mid 20, 5)$, $P(0 \mid 16, 4)$, $P(1 \mid 12, 4)$ and $P(2 \mid 9, 3)$. Without bothering
to make the distributions continuous, we find that the numerical values
of these tails are 0.057, 1.000, 0.125, and 0.000152, and the combined
tail-area probability, by Fisher's method, is 1/1700. If we had made the
distributions continuous, this last method would, I think, be the most
reliable of the non-Bayesian methods. As it is, they all give much the
same result, and the most reliable of them is the *ad hoc* method for
which the tail-area was 1/2300. In this example, we can obtain the Bayes
factor approximately by taking the reciprocal of the tail-area probability
and dividing by about 20. This is consistent with my impression that this
correction factor is usually in the range (10/3, 30) (see also, Good[34]).
On the whole the correction factor seems to be larger for very small
tail-area probabilities than for moderate ones.

It will be noticed that in the numerical example an examination of
the form of the Type II likelihood, as a function of k (see Table 5.1),
would not have led to the rejection of either the Perks or the Jeffreys
values of k. It can easily be seen however that if N is very large and if
the ratios n_i/N are close to $1/t$, but significantly different from $1/t$, we
would be forced to assume that k was large. Conversely, if one of the
n_i's were very small in a very large sample, we would be forced to assume
that k is very small. This shows once again that the Type II initial
distribution cannot strictly be a symmetrical Dirichlet distribution and
that, if this distribution is to be used at all, we must be prepared to
assume that k has a Type III distribution. One defense for such a
procedure is that W. E. Johnson's sufficiency postulate is not entirely
unreasonable, and our procedure is an improvement of it. The Type III
distribution may be described by the philosophically even more neutral
name of *weighting system*.

It is a familiar principle that we must be prepared to reject a null
hypothesis of Type I when the evidence against it is great. We are
claiming that the same principle applies to null hypotheses of Type II,
although the sample required to reject a Type II hypothesis will often
need to be much larger than that required to reject one of Type I in the
same context. The thesis could be extended to Type III hypotheses and
beyond, but it would seldom need to be so extended.

Just as it is convenient to assume the approximate truth of a Type I
hypothesis, when it has formal simplicity and is consistent with the
evidence, the same applies to Type II and Type III hypotheses, whether
they be physical, logical, subjective, or psychological.

We have discussed in this chapter significance tests for the hypothesis $k = \infty$. More generally, the factor against the simple statistical hypothesis ($p_i = q_i$), provided by the sample (n_i), when the alternative is that the Type I population has been selected by means of a Type II Dirichlet distribution of parameters (k_i) is (see Good,[33] p. 862)

$$\frac{\Gamma(\sum k_i) \prod \Gamma(n_i + k_i)}{\Gamma(N + \sum k_i) \prod \{\Gamma(k_i) q_i^{n_i}\}} \tag{5.22}$$

We might take $k_i = k$ or $k_i = ktq_i$, depending on whether we assume, for the non-null hypothesis, $\mathscr{E}_{II}(p_i) = 1/t$ or q_i. If k is unknown, then the maximum factor might be usable, or a Type III initial distribution for k might be assumed, as before.

A more general problem to which we referred earlier, will now be discussed in the spirit of the present philosophy.

5.4 Discrimination Between Two Populations When the Basic Samples Are Finite or Infinite

Suppose that we have two *basic samples* (r_i) and (s_i) of sample sizes R and S and a *nonbasic sample* (n_i) of size N. (The notation is the same as that at the beginning of this chapter.) The two basic samples come from distinct populations (p_i) and (q_i), and it is known that the nonbasic sample comes from one or the other of these two populations also. The problem is to obtain a Bayes factor in favor of H_p as compared with H_q, these being names for the two (Type I) hypotheses. We know that it would be dangerous to compute the log-factor with the aid of the maximum-likelihood estimates r_i/R and s_i/S of p_i and q_i. A not very Bayesian statistician might agree to flatten the basic samples, using the flattening constants suggested at the beginning of this chapter. But we know that the use of a flattening constant is equivalent to the assumption of symmetrical Dirichlet Type II initial distributions. Let the flattening constants, that is, the Dirichlet parameters, be k and k', respectively. Let E denote the observed event that the three samples have the frequency counts (r_i), (s_i), and (n_i). Then it is easily seen that the probability of this event, given H_p, is

$P(E \mid H_p, k, k')$

$$= \frac{R! S! N! \Gamma(tk) \Gamma(tk') \prod \Gamma(s_i + k') \prod \Gamma(r_i + n_i + k)}{\prod (r_i! s_i! n_i!)(\Gamma(k))^t (\Gamma(k'))^t \Gamma(S + tk') \Gamma(R + N + tk)} \tag{5.23}$$

and the probability $P(E \mid H_q, k, k')$ is given by a similar expression.

In order to compute a Bayes factor in favor of H_p as against H_q, it is necessary to assume a Type III joint distribution $G(k, k')$ for (k, k'). We then have

$$F(H_p/H_q:E) = \frac{\iint P(E \mid H_p, k, k') \, d^2G(k, k')}{\iint P(E \mid H_q, k, k') \, d^2G(k, k')} \qquad (5.24)$$

As an example, let us suppose that the two Type I populations (p_i) and (q_i) are obtained by (Type II) sampling with symmetrical Dirichlet Type II initial distributions having parameters k and k', where k and k' independently have the (Type III) distribution specified in Formulas 5.20. Then

$$F = F(H_p/H_q:E) = \frac{\Psi(r + n)\Psi(s)}{\Psi(s + n)\Psi(r)} \qquad (5.25)$$

where

$$\Psi(m) = \int_0^\infty \psi(m, k)\phi(k) \, dk \qquad (5.26)$$

where

$$\psi(m, k) = \frac{\Gamma(tk) \prod \Gamma(m_i + k)}{\Gamma(k)^t \Gamma(M + tk)}, \qquad M = \sum m_i \qquad (5.27)$$

We note for future reference, that $\psi(m, k) \to t^{-M}$ when $k \to \infty$. Also, if $M \to \infty$ and $m_i/M \to \mu_i$, then the value of k that maximizes $\psi(m, k)$ is equal to the value that maximizes

$$\Gamma(tk)\Gamma(k)^{-t} \prod \mu_i^k$$

and, if t is not very small (that is, large enough for the application of Stirling's formula), this value of k is approximately

$$\frac{-\frac{1}{2}(t - 1)}{\sum \log (t\mu_i)} \qquad (5.28)$$

If the μ_i's are close to $1/t$, this Type II maximum-likelihood value of k is approximately equal to the value given by Equation 5.11 (with $p_i = \mu_i$).

In most applications, the Type III distribution of (k, k') will be vague, and the factor calculated with its aid will not be better than a reasonable approximation to the "true" Bayes factor (if there is one). It might benefit relationships with other statisticians if we use a formula that makes no explicit reference to a Type III distribution. Three candidates (Equations 5.29, 5.30, and 5.32) are the Type II likelihood ratios

$$\begin{aligned} G_1 = G_1(H_p/H_q:E) &= \frac{\max_{k,k'} P(E \mid H_p, k, k')}{\max_{k,k'} P(E \mid H_q, k, k')} \\ &= \frac{\psi(s, \hat{k}_s)\psi(r + n, \hat{k}_{r+n})}{\psi(r, \hat{k}_r)\psi(s + n, \hat{k}_{s+n})} \end{aligned} \qquad (5.29)$$

where $\hat{k}_m$ maximizes $\psi(\mathbf{m}, k)$; and

$$G_2 = G_2(H_p/H_q : E) = \frac{\max_k P(E \mid H_p, k = k')}{\max_k P(E \mid H_q, k = k')} = \frac{\beta(\mathbf{s}, \mathbf{r} + \mathbf{n})}{\beta(\mathbf{r}, \mathbf{s} + \mathbf{n})} \tag{5.30}$$

where

$$\beta(\mathbf{l}, \mathbf{m}) = \max_k \frac{\Gamma(tk)^2}{\Gamma(k)^{2t}} \cdot \frac{\prod \Gamma(l_i + k)\Gamma(m_i + k)}{\Gamma(kt + \sum l_i)\Gamma(kt + \sum m_i)} \tag{5.31}$$

and finally a Type II likelihood ratio, adjusted for the curvatures of the likelihood curves or surfaces,

$$G_3 = G_3(H_p/H_q : E) = \frac{\omega(\mathbf{s}, \hat{k}_s)\omega(\mathbf{r} + \mathbf{n}, \hat{k}_{r+n})}{\omega(\mathbf{r}, \hat{k}_r)\omega(\mathbf{s} + \mathbf{n}, \hat{k}_{s+n})} \tag{5.32}$$

where

$$\omega(\mathbf{m}, \hat{k}_m) = \frac{\psi(\mathbf{m}, \hat{k}_m)}{\{d^2/dk^2 \log 1/[\psi(\mathbf{m}, k)]\}_{k=\hat{k}_m}^{1/2}} \tag{5.33}$$

The effect of "correcting for curvature" is to replace a maximum likelihood by a number proportional to the area (or volume) under the likelihood curve (or surface) in the neighborhood of the maximum. The reason for doing this is that the result will be roughly proportional to the integral of the likelihood weighted with a rather uniform Type III initial distribution for k, uniform, that is, in the neighborhood of the maximum-likelihood value of k. (We are considering, not advocating, this suggestion.)

Note that, in the Type II likelihood ratios, there are the same number of parameters, with respect to which maximization is done, in the numerator and in the denominator. If it were not for this, we would not expect the likelihood ratios to give a reasonable approximation to a Bayes factor.

As a first example, we take the following frequencies:

i	1	2	3	4	5		
r_i	17	5	16	3	9	$R = 50$	$(x^2 = 16)$
s_i	10	2	4	16	18	$S = 50$	$(x^2 = 20)$
n_i	1	3	2	4	2	$N = 12$	

In a second example, n_5 was changed from 2 to 7 (and N, therefore, from 12 to 17). In the first example, Formulas 5.25, 5.29, 5.30, and 5.32 gave (with the help of a FORTRAN program)

$$F = F(H_p/H_q : E) = 1/3.6, \quad G_1 = 1/2.5, \quad G_2 = 1/4.75, \quad G_3 = 1/1.8$$

and, in the second example,

$$F = F(H_p/H_q : E) = 1/35, G_1 = 1/23, G_2 = 1/45, G_3 = 1/14$$

In both examples, G_1 and G_2 are adequately close to F, and the geometric means $(G_1 G_2)^{1/2}$ are very close to F, proportionately. The closeness of the geometric mean might be a fluke, and it would be interesting to look at several further examples. The Type II likelihood ratio, corrected for curvature, agrees somewhat *less* well than the uncorrected ratios.

The Type II maximum-likelihood values of k, say $\hat{k}$, for the frequency counts (r_i), (s_i), $(r_i + n_i)$, and $(s_i + n_i)$, with the values of k corresponding to the Formula 5.9 in parentheses, say $\hat{k}*$, are, for the first example, 2.85 (3.3), 1.9 (2.5), 9.4 (7.8), and 3.4 (3.3). These values of $\hat{k}$ and $\hat{k}*$ must be regarded as approximately equal, since the values of

$$\frac{\psi(\mathbf{m}, \hat{k})}{\psi(\mathbf{m}, \hat{k}*)}$$

in these four cases are all very close to 1, being in fact

$$1.014, \ 1.064, \ 1.014, \ \text{and} \ 1.002$$

It seems that one might as well use $\hat{k}*$ instead of $\hat{k}$, when calculating the Type II likelihood ratios. On the other hand, the values 0.2 and 0.5 for k, as provided by the invariance theories of Perks and Jeffreys, respectively, give

$$\frac{\psi(\mathbf{m}, \hat{k})}{\psi(\mathbf{m}, 0.2)} = 62, 29, 411, 10$$

$$\frac{\psi(\mathbf{m}, \hat{k})}{\psi(\mathbf{m}, 0.5)} = 6, 3.5, 35, 9$$

so that, in most of these cases, the invariance theories would be "quantitatively inconsistent" with Type II maximum likelihood. Nevertheless, the factors in favor of H_p, if Perks' value for k is used, and if Jeffreys' value is used, are, respectively, in the two examples,

$$F_{\text{Perks}}(H_p/H_q : E) = \frac{\psi(\mathbf{s}, 1/5)\psi(\mathbf{r} + \mathbf{n}, 1/5)}{\psi(\mathbf{r}, 1/5)\psi(\mathbf{s} + \mathbf{n}, 1/5)} = 1/5.0 \text{ and } 1/56$$

$$F_{\text{Jeffreys}}(H_p/H_q : E) = \frac{\psi(\mathbf{s}, 1/2)\psi(\mathbf{r} + \mathbf{n}, 1/2)}{\psi(\mathbf{r}, 1/2)\psi(\mathbf{s} + \mathbf{n}, 1/2)} = 1/5.1 \text{ and } 1/58$$

These are extremely close to each other, and quite close to G_2 in both examples. For purposes of discrimination between H_p and H_q, the Bayes factors seem to be insensitive to the selection of k or the selection of a Type III distribution for k, although the factor in favor of a particular value of k, as against another one, might be large. One model can have a much larger likelihood than another, and yet both can give much the same Bayes factor in favor of H_p. *By virtue of this phenomenon, the amount of self-interrogation necessary for statistical inference is not as great as it would otherwise need to be.*

Under what circumstances, and by what arguments, could we reject the invariance theories? It is not enough that they are quantitatively inconsistent with Type II maximum likelihood. What would make us reject say the value $k = 0.2$ would be the observation that a *wide* range of values of k were all of much higher (Type II) likelihood than $k = 0.2$. But what would we mean by "wide"? Surely we must mean wide enough so that the total initial probability that k lies in this range is by no means negligible. If we had a fixed idea that $k = 0.2$, then no range could be wide enough to reject this value. I suggest that no one really has such a fixed idea; Perks and Jeffreys both put forward their invariance theories somewhat tentatively. In order to reject the invariance theories, one can use the device of imaginary results; it is by no means necessary to consider only the results of actual experiments. In fact it might be a mistake to do so, since the invariance theories are supposed to apply when there are no additional facts to obscure the issue, as there usually are in any real problem. Accordingly, we first take the example of a five-category multinomial sample (18, 8, 18, 7, 11). We find in this example that every value of $k > 1.5$ has more than a hundred times the likelihood of the value $k = 0.2$, and every value $k > 1.7$ has more than ten times the likelihood of the value $k = 0.5$. One can also express the argument against the invariance theories in terms of the population frequencies (p_i). When the repeat rate $\sum p_i^2$ is close to $1/t$, a wide range of values of k will have much more Type II likelihood than either of the values $k = 1/t$ and $k = 1/2$. It does not seem reasonable to me that the class of Type I populations for which the repeat rate is fairly close to $1/t$ should as a general rule be jointly assigned a very low initial probability. This is of course a subjective opinion, but one, I think, that would be shared by many statisticians.

This last argument can be made more quantitative in the following manner: Let $\mathcal{H}_k$ denote the Type II hypothesis that the population (p_i) is selected from a superpopulation with the Type II initial distribution of the symmetrical Dirichlet form with parameter k. If we take a large enough Type I sample (n_i), we will be able to estimate the p_i's fairly accurately, and we will find that the factor in favor of a particular value of k as against the value $k = 1/2$ is

$$x(k) = \frac{\Gamma(kt)\pi^{t/2}}{(\Gamma(k))^t \Gamma(\tfrac{1}{2}t)} \prod p_i^{k-1/2} \tag{5.34}$$

When t and k are both large, and when $\rho = \sum p_i^2$ is close to $1/t$,

$$x(k) \sim (2k)^{1/2}(ek)^{t/2} \exp\{-\tfrac{1}{2}t(t\rho - 1)(k - \tfrac{1}{2})\} \qquad (5.35)$$

If the initial (subjective) probability

$$P\left(1 < t\rho < 1 + \frac{1}{t}\right) > t^{-c}$$

for any constant c, we are virtually forced to reject the Type II hypothesis $\mathscr{H}_{1/2}$, and more generally to reject any hypothesis of the form $k < k_0$ if t is large enough. In other words, it does not seem reasonable to assign zero Type III probability to the hypothesis $k > k_0$, for any k_0, when t is large enough. Roughly speaking, what this amounts to is that we should assume a Type III distribution for k rather than assume some definite value for k.

5.5   Special Cases of the Discrimination Problem

1. *Zero nonbasic sample.* When $N = 0$, the Bayes factor $F(H_p/H_q:E)$ reduces to 1, whatever Type III initial distribution for (k, k') is assumed, and also the Type II likelihood ratio reduces to 1.

2. *Both basic samples infinite.* In this case, the log-factor, for all initial Type III distributions, and also $\log G_1$, $\log G_2$, and $\log G_3$ reduce to $\sum n_i \log(p_i/q_i)$, as they should (see the beginning of this chapter).

3. *Significance test for whether a sample* (n_i) *came from a specified (Type I) population* (p_i). Here we take R as infinite and S as zero, but it is simpler to work *de novo* instead of using the formulas for finite R. The factor against the null hypothesis is (see Good,[33] p. 862)

$$F = \prod p_i^{-n_i} \int_0^\infty \psi(\mathbf{n}, \mathbf{k})\phi(k) \, dk \qquad (5.36)$$

TABLE 5.2

Example	i:	1	2	3	4	5	
	p_i:	0.34	0.10	0.32	0.06	0.18	
(a)	n_i:	1	3	2	4	2	$N = 12$
(b)	n_i:	1	3	2	4	7	$N = 17$
(c)	n_i:	34	10	32	6	18	$N = 100$
(d)	n_i:	17	5	16	3	9	$N = 50$
(e)	n_i:	22	6	10	3	9	$N = 50$

where $\phi(k)$ is the Type III initial density of k. As examples, take $\phi(k)$ as in Equations 5.20, and p_i and n_i as in Table 5.2. For these five examples, the values of

$$\chi_0^2 = \sum \frac{(n_i - Np_i)^2}{Np_i} \tag{5.37}$$

and of the likelihood-ratio statistic

$$\mu = 2 \sum n_r \log_e n_r - 2 \sum n_i \log_e p_i - 2N \log_e N \tag{5.38}$$

were calculated, together with their tail-area probabilities $P(\chi_0^2)$ and $P(\mu)$, on the assumption that the tabular chi-squared approximation could be used. The results, and the values of F, are shown in Table 5.3.

TABLE 5.3

Example	χ_0^2	μ	$P(\chi_0^2)$	$P(\mu)$	F
(a)	20.9	13.5	1/3000	1/109	65
(b)	20.9	18.4	1/3000	1/980	260
(c)	0	0	1	1	1/1300
(d)	0	0	1	1	1/310
(e)	3.9	4.1	1/2.4	1/2.5	1/42

(In Table 5.3, F was calculated with a FORTRAN program, the likelihood-ratio statistic μ was calculated with the aid of the table of $2n \log_e n$ given in the appendix of Ku[3],[76] and the tails of the incomplete gamma function with the aid of Karl Pearson[92]: $P(\chi^2 > \chi_0^2) = 1 - I(\chi_0^2/\sqrt{2d}, \frac{1}{2}d - 1)$ in Pearson's notation $I(u, p)$, where d is the number of degrees of freedom.)

Note that the factor in favor of the hypothesis H_0 is greater in Example (c) than in Example (d), but χ_0^2 and μ are 0 for both examples and so would not in themselves lead to any distinction. It is intuitively obvious that the evidence in favor of H_0 is greater in Example (c) than in Example (d), since the size of sample is greater in Example (c).

4. *Significance test for whether two samples* (r_i) *and* (n_i) *arose from the same (Type I) population.* Here $S = 0$, and the factor against the null hypothesis reduces to

$$F' = \frac{\Psi(\mathbf{r})\Psi(\mathbf{n})}{\Psi(\mathbf{r} + \mathbf{n})} \tag{5.39}$$

which is symmetrical in the two samples and is equal to unity when either sample is of zero size, as it should be. The Type II likelihood-ratio statistics would also be of interest, but I have not calculated numerical examples of them.

The problem of testing whether two samples arose from the same population is of course identical with that of testing for independence (no association) in a $2 \times t$ contingency table, when the two row sums are considered to give no evidence, or a negligible amount. In order that the tail should not wag the dog, we discuss this topic in a separate chapter.

In this chapter we have relied on initial Type II distributions of Dirichlet form or weighted averages of such distributions. It should be noted however that if, for multiple sampling, we can accept *any* definite formula for $P(j \mid (n_i))$, the probability that the next letter will be j when the sample so far is (n_i), then it is possible to write down the likelihoods of all the hypotheses considered in terms of this formula. It is thus not essential to base all our results on a modification of Johnson's sufficiency postulate. But I believe it is usually adequate to do so when samples are not extremely large.

Since we have not accepted Johnson's sufficiency postulate, the technique of assuming a linear combination of symmetrical Dirichlet distributions cannot be fully justified by exact logic. What is claimed is that it is a better assumption than that of any single Dirichlet distribution, and gives a good and convenient Bayesian solution to the problem of multinomial discrimination, and to some problems concerning contingency tables. As Jeffreys said of the brain as a thinking machine, it is not perfect but is the only one available.

6. Tests for Independence in Contingency Tables

We begin this chapter by continuing the discussion of the example at the end of Chapter 5. As numerical examples, we take five 2×5 contingency tables determined by the five pairs of frequency counts $[a, a]$, $[a, d]$, $[a, e]$, $[d, d]$, and $[d, e]$, where the counts (a), (d), and (e) are, as in Table 5.2, $(1, 3, 2, 4, 2)$, $(17, 5, 16, 3, 9)$, and $(22, 6, 10, 3, 9)$, respectively. Table 6.1 gives the values of the log likelihood ratio

$$\mu = \sum 2r_i \log_e r_i + \sum 2n_i \log_e n_i + 2(R + N) \log_e (R + N)$$
$$- 2R \log_e R - 2N \log_e N - \sum 2(r_i + n_i) \log_e (r_i + n_i) \qquad (6.1)$$

TABLE 6.1

Example	μ	$P(\mu)$	F'	F''	f	f'	f''
$[a, a]$	0	1	1/10	1/12	1/2.8	1/6	1/7
$[a, d]$	10.07	1/27	13	1/1.4	4.4	9	1.5
$[a, e]$	10.44	1/31	18	1/1.1	5.4	9	1.8
$[d, d]$	0	1	1/77	1/74	1/31	1/74	1/57
$[d, e]$	2.13	1/1.4	1/29	1/29	1/11	1/33	1/21

the corresponding tail-area probability on the assumption that the tabular chi-squared tables give an adequate approximation, and various Bayes factors, explained later.

Table 6.1 exemplifies the fact that Bayesian methods can provide measures for the evidence in favor of a null hypothesis, for which purpose non-Bayesian methods are not as well tailored. (This is exemplified for μ. The same is true for χ^2.) Non-Bayesians often say that null hypotheses can be rejected but not confirmed. The Bayesian believes that they can be confirmed, at least as approximations to the truth, while agreeing that, for moderate-sized samples, the expected weight of evidence in favor of a true null hypothesis is usually smaller than that against a false one. This comment has a bearing on the philosophy of all sciences, since many scientific theories can be regarded as null hypotheses.

In order to explain the meanings of F', F'', f, f', and f'', we must first say something about the general two-dimensional contingency table (n_{ij}), with s rows and t columns ($i = 1, 2, \cdots, s; j = 1, 2, \cdots, t$). The meaning of the symbol s should not be confused with that in Chapter 5, which we drop now.

If the row totals $(n_{i.})$ are given and the contingency table is then built up by sampling, we would surely judge that the row totals contribute no evidence in favor of the null hypothesis of independence (no association) in the contingency table. Then an extension of the previous theory (see Equation 5.39) leads to the factor against the null hypothesis,

$$F' = \frac{\prod_i \Psi((n_{ij})_{j=1,2,\cdots,t})}{\Psi((n_{.j})_{j=1,2,\cdots,t})} \qquad (6.2)$$

where the argument of Ψ in the denominator is the vector consisting of the column totals, and the argument of the ith factor in the numerator is the vector consisting of the ith row of the contingency table. The function Ψ is defined by Equation 5.26, where $\phi(k)$ is the Type III density of k. In the numerical examples we continue to assume that ϕ is defined by Equations 5.20.

If the column totals are first fixed and are assumed to contribute no evidence in themselves, then the factor would be given by interchanging rows and columns. The factors F'' are calculated for the five 2×5 contingency tables, on the latter assumption. It might seem surprising that for the contingency tables $[a, d]$ and $[a, e]$, for which one would be inclined to reject the null hypothesis, F'' is close to 1. The reason F'' is so small, is that, five times, the theory treats a column (two cells) of the contingency table as if it were obtained in a manner symmetrical with respect to rows 1 and 2. If we think of the columns as arriving one at a time, the Type II expectations of the two cell entries are taken as equal for all five columns. Although the assumed model is entirely possible (remember that all our probability distributions *could* be physical ones), in most circumstances it would not be reasonable: as we proceed from one column to the next, we ought to learn by experience and should be prepared to reject the symmetrical Type II distribution. (See also Appendix E.) For this reason, in the Bayesian analysis of a $2 \times t$ table ($t < 2$), it will nearly always be more sensible to regard the two row totals as given, and as contributing negligible evidence, than to regard the column totals as given. More generally, the formula for F' (Equation 6.2), although of some theoretical interest, is barely applicable when $s > 2$.

Further numerical support for this argument can be obtained from the Bayesian models considered by Good,[27] pp. 98–101. It was there pointed out that the factors in favor or against the null hypothesis can be written down explicitly *if* various Bayes postulates of uniform distribution are made. The factors f, f', f'' listed in Table 6.1, are the factors against the null hypothesis if (1) the marginal totals are both regarded as relevant evidence by themselves, (2) the row totals are regarded as irrelevant, and (3) the column totals are regarded as irrelevant. In a footnote in that book, added in proof, it was tentatively suggested that $f'f''/f$ could reasonably be taken as the factor if neither the row columns nor the column totals are regarded as relevant, but I now doubt this. It will be seen from the numerical results that f' is not far from F' and f'' is not far from F''. Thus the precise Bayesian model is apparently not responsible for the big difference between F' and F'', for the tables $[a, d]$ and $[a, e]$.

6.1 Row and Column Totals Irrelevant: a More Acceptable Approach

We should like a new Bayesian model for the case where both the row and the column totals are regarded as contributing negligible evidence, even when both lots of totals are given. One possible idea is to assume that, given the non-null hypothesis, all contingency tables subject to the given marginal totals are equiprobable. Then each of these probabilities would be equal to

$$\frac{1}{\mathscr{C}(\prod x_i^{n_i \cdot} y_j^{n \cdot j}) \prod (1 - x_i y_j)^{-1}} \tag{6.3}$$

where $\mathscr{C}(\dots)$ denotes "coefficient of ... in." If this coefficient could be explicitly extracted, one could get a factor against the null hypothesis, since, given the null hypothesis, the conditional probability of the table is given by Yates' formula (see, for example, Mood,[86] p. 278),

$$\frac{\prod n_{i\cdot}! n_{\cdot j}!}{N! \prod n_{ij}!} \tag{6.4}$$

For a 2×2 table, this suggestion turns out to give the result of Jeffreys,[63] p. 260, except that his philosophy is different and his result arises only as an approximation. It may be noted that, for a 2×2 table, the coefficient mentioned here reduces to 1 plus the least of the four marginal totals. The method is also in the spirit of the method suggested for multinomial distributions by Jeffreys,[63] p. 67. It is *not* in the spirit of the Type III distribution for k, advocated in the present work.

The following more general method *is* in the spirit of the present work. Imagine first that we know only N, and assume that the st probabilities of the cells of the contingency table have, on the non-null hypothesis $\bar{H}_0$, a Type II symmetrical Dirichlet distribution with parameter k, where k itself has a Type III density $\phi(k)$. Then the probability of the observed event (n_{ij}) is

$$P((n_{ij}) \mid \bar{H}_0, N) = \frac{N!}{\prod n_{ij}!} \int_0^\infty \frac{\Gamma(stk) \prod \Gamma(n_{ij} + k)}{\Gamma(k)^{st} \Gamma(N + stk)} \phi(k) \, dk$$

The conditional probability of (n_{ij}), conditional on the marginal totals $(n_{i\cdot})$ and $(n_{\cdot j})$, is

$$P((n_{ij}) \mid \bar{H}_0, (n_{i\cdot}), (n_{\cdot j})) = \frac{P((n_{ij}) \mid \bar{H}_0, N)}{\sum P((n_{ij}) \mid \bar{H}_0, N)} \tag{6.5}$$

where the summation is over all values of n_{ij} for which

$$\sum_j n_{ij} = n_{i\cdot}, \qquad \sum_i n_{ij} = n_{\cdot j} \quad (i = 1, 2, \cdots, s; j = 1, 2, \cdots, t)$$

It follows that

$$P((n_{ij}) \mid \bar{H}_0, (n_{i\cdot}), (n_{\cdot j}))$$
$$= \frac{\int_0^\infty \frac{\Gamma(stk) \prod \Gamma(n_{ij} + k)/\Gamma(n_{ij} + 1)}{\Gamma(k)^{st} \Gamma(N + stk)} \phi(k) \, dk}{\mathscr{C}(\prod x_i^{n_{i\cdot}} \cdot y_j^{n_{\cdot j}}) \int_0^\infty \frac{\Gamma(stk) \prod (1 - x_i y_j)^{-k}}{\Gamma(N + stk)} \phi(k) \, dk} \tag{6.6}$$

On dividing by Formula 6.4, one gets the factor against the null hypothesis H_0. (Of course, $\mathscr{C}$ and $\int$ commute.)

In the special case $\phi(k) = \delta(k - 1)$, a Dirac function, this method reduces to the one mentioned earlier. More generally, if a symmetrical Dirichlet distribution is assumed, with parameter k, where k is not necessarily equal to 1, the factor against the null hypothesis provided by (n_{ij}), when the marginal totals are given and are assumed to convey no evidence in themselves, is

$$F_k(\overline{H}_0/H_0) = F(\overline{H}_0/H_0:(n_{ij}) \mid (n_{i.}), (n_{.j}))$$

$$= \frac{N! \prod \Gamma(n_{ij} + k)}{\Gamma(k)^{st} \prod n_{i.}! n_{.j}! \mathscr{C}(\prod x_i^{n_{i.}} y_j^{n_{.j}}) \prod (1 - x_i y_j)^{-k}} \quad (6.7)$$

(As mentioned on page 32, the oblique stroke denotes "as against," the colon "provided by," the vertical stroke "given.")

For a 2×2 contingency table,

$$F_k(\overline{H}_0/H_0) = \frac{N! \prod \Gamma(n_{ij} + k)}{S_k \prod n_{i.}! n_{.j}!} \quad (6.8)$$

where

$$S_k = \sum_{v=0}^{\mu} \frac{\Gamma(\mu + k - v)\Gamma(k + v)\Gamma(a + k + v)\Gamma(b + k - v)}{(\mu - v)! v! (a + v)! (b - v)!} \quad (6.9)$$

where

$$\begin{bmatrix} \mu, 0 \\ a, b \end{bmatrix} \quad (6.10)$$

is a contingency table consistent with the marginal totals. In particular, $S_1 = \mu + 1$.

As an example, consider the 2×2 contingency table concerned with the criminality of the twins of criminals (Fisher,[22] p. 94, quoting from Lange;[77] and Jeffreys,[63] p. 265):

	Monozygotic	Dizygotic
Convicted	10	2
Not Convicted	3	15

The tail-area on Fisher's "exact test" is 1/2150. By Jeffreys' test, the factor against the null hypothesis is 171, which he interprets as odds since, as is his wont, he takes the initial odds of the null hypothesis as 1. The factors $F_k(\overline{H}_0/H_0)$, for various values of k, are listed in Table 6.2. I think it will be agreed that, with any reasonable weighting function (Type III distribution), the factor against the null hypothesis should be somewhat less, but not a great deal less, than Jeffreys' value. If the density function $\phi(k)$ defined by Equations 5.20 is used, the weighted average of the factors $F_k(\overline{H}_0/H_0)$ is about 120. But some people might prefer to base their judgment on the list of factors in Table 6.2, without any explicit assumption concerning their relative weights. (Compare the preference often shown by non-Bayesians for looking at the set of Type I likelihoods rather than weighting them with an initial distribution.)

TABLE 6.2

k	$F_k(\overline{H}_0/H_0)$		k	$F_k(\overline{H}_0/H_0)$
0.2	78		1.6	141
0.4	134		1.8	130
0.6	163		2.0	119
0.8	173		3.0	79
1.0	171		4.0	54
1.2	163		5.0	39
1.4	153			

For comparison, the factors F' and F'', calculated as in Table 6.1, that is, first with the row totals regarded as given, and second with the column totals regarded as given, are $F' = 163$, $F'' = 168$. For this 2×2 contingency table, at any rate, the various Bayesian models lead to much the same factor against the null hypothesis.

The approach discussed here seems to me to provide a more satisfactory basis for the Bayesian treatment of a 2×2 contingency tables than was available before.

For arbitrary tables with two rows, the factor F', for which the row totals are regarded as given and as contributing negligible evidence, seems to be adequate. For tables with both s and t greater than 2, a tabulation of $F_k(\overline{H}_0/H_0)$ would presumably supply an adequate picture for judging whether the null hypothesis of independence should be rejected, if the initial probability of this hypothesis and the utility losses of rejecting and accepting it are taken into account. But unfortunately the calculations are liable to be too heavy. Consequently we are liable to be forced to use a non-Bayesian method in this case, such as the use of χ^2 or the likelihood ratio, μ. The asymptotic distribution of these statistics, the "tabular chi-squared" distribution, is unreliable when the cell expectations are very small, such as less than unity. In such cases at any rate the exact expectation and variance of χ^2 can be calculated, assuming the null hypothesis. For very sparse contingency tables, $R = \sum n_{ij}(n_{ij} - 1)/2$ has approximately a Poisson distribution to a good approximation. (This follows from von Mises.[85]) If the calculation of the factors $F_k(\overline{H}_0/H_0)$ could be performed for large contingency tables, they would be especially interesting for tables too sparse for the tabular chi-squared distribution to be reliable, and not sparse enough for the Poisson distribution of $\sum n_{ij}(n_{ij} - 1)/2$ to be reliable.

It is not yet known in quantitative terms what is meant here by "too sparse" and "not sparse enough," and this ignorance would make the calculation of $F_k(\overline{H}_0/H_0)$ more useful.

6.2 Example of an *Ad Hoc* Method for a Large Sparse Table

In the absence of a practicable method for doing these calculations, it is advisable in practice to be prepared to make use of *ad hoc* methods. This is exemplified by the data concerned with amino acid allele pairs in Table 6.3, kindly supplied by Dr. Richard V. Eck. (For the background, see Eck.[18]) For genetical reasons, the contingency table can be folded about its main diagonal, and the cell entries are listed in Table 6.3. The dots represent omitted diagonal entries. The twenty rows

TABLE 6.3

.1030513103000181010	28
.000000001300020010	8
.00000000000000000000	0
.041110000000000000	10
.000000000000000	0
.22001300020010	20
.0101000000000	6
.000011010010	10
.00000000000	3
.0001001030	5
.100020000	9
.00020010	10
.0000010	2
.000000	2
.10000	2
.1010	20
.010	4
.00	0
.1	12
	1
	152 = 2N

correspond to amino acids. We shall not here discuss the genetical meaning of this folded contingency table. The marginal totals are those of "bent rows," and *their* total is $2N$, where N is the sample size.

On the null hypothesis of independence, the conditional probability of the table, given its marginal totals (n_i), can be shown to be (see Appendix C, and Ishii[60])

$$\frac{2^{N - \Sigma n_{ii}} N! \prod_{i \leq j} n_i!}{(2N)! \prod_{i, j} m_{ij}!} \tag{6.11}$$

where m_{ij} are the numbers entered in the cells of the folded table $(m_{ij} = n_{ij} + n_{ji})$. The expectation of R is

$$\mathscr{E}(R \mid (n_i)) = \frac{(\sum n_i^2 - \sum n_i)^2 - \sum (n_i^2 - n_i)^2}{4(2N - 1)(2N - 3)} \tag{6.12}$$

By making various iterative adjustments to allow for the fact that the leading diagonal is not entered, it can be shown that the excess of the observed value of R over its expectation on the null hypothesis can be entirely attributed to the large entry $m_{1,16} = 8$ in the top row. Now

$$\mathscr{E}(m_{ij}) = \frac{n_i n_j}{2N - 1} \qquad (i < j) \tag{6.13}$$

$$\mathscr{E}(m_{ij}^2 - m_{ij}) = \frac{n_i(n_i - 1)n_j(n_j - 1)}{(2N - 1)(2N - 3)} \tag{6.14}$$

Therefore, $\mathscr{E}(m_{1,16}) = 3.71$, var $(m_{1,16}) = 2.72$. By fitting a binomial distribution, we find that the probability that $m_{1,16} \geq 8$ is somewhat greater than 1/100. But the contingency table contains more than 100 cells. Therefore, it contributes evidence in *favor* of the null hypothesis if anything. Matters would have been quite different if the 8 had occurred say in cell (2, 7). In genetical terms this means that the contingency table certainly supplies no evidence against the hypothesis of a nonoverlapping genetic code.

   This example has been presented here in order to show that none of the methods mentioned earlier is necessarily adequate for the discussion of a practical problem. In our theories, we rightly search for unification, but real life is both complicated and short, and we make no mockery of honest adhockery.

7. Large Pure Contingency Tables

Let (n_{ij}) $(i = 1, 2, \cdots, s; j = 1, 2, \cdots, t)$ be a large pure contingency table, where "large" means having at least five rows and five columns and "pure" means that there is no natural ordering of the rows or of the columns. In practice, large contingency tables are rarely pure, but purity might sometimes be an adequate model. Suppose that the hypothesis of no association between the rows and columns has been definitely rejected. *We consider the problem of estimating the population probability p_{ij} of cell (i, j).* When n_{ij} is small, our problem is one of estimating a probability from an effectively small sample. If the problem were regarded as a special case of that of sampling from a multinomial population, then the information that we have a two-way classification would be ignored.

This problem was apparently first considered, for application to census work, by Deming and Stephan [16] by regarding the marginal population frequencies $p_{i.}$ and $p_{.j}$ as known, and minimizing

$$\sum \frac{(p_{ij} - n_{ij}/N)^2}{n_{ij}}$$

subject to the constraints $\sum_j p_{ij} = p_{i.}$, $\sum_i p_{ij} = p_{.j}$. (They did not explicitly affirm the purity of the contingency table.) Their method leads to $p_{ij} = 0$, when $n_{ij} = 0$, so in spite of the convenience and usefulness of their method, it is not satisfactory in this case. (See also the discussion of multidimensional tables in Chapter 9.)

Being interested in the estimation of the probabilities of events that have never occurred, Good[31] tackled the problem in the following manner. Two large contingency tables were taken from real life. One concerned the occupations of fathers and sons, for which the diagonal is special, and the other concerned occupations and causes of death. It was found that the association factor $p_{ij}/(p_{i.}p_{.j})$ appeared to have a lognormal distribution over the st cells of each table. (In other words the "mutual information" had a normal distribution.) By estimating the distribution from the contingency table itself, and treating it as an initial, or rather a *semi-initial*, distribution for any given cell, it was possible to derive a final distribution of p_{ij} from the three observations n_{ij}, $n_{i.}$, and $n_{.j}$. The result was expressed in terms of the "after effect function" (*Nachwirkungsfunktion*), which was originally tabulated for an electrodynamical application (Wagner[108]). In this manner, Type II expectations and variances were estimated for p_{ij}. It was found that much the same results were obtained if the association factor was assumed to have a Pearson distribution, of Pearson's third type, even though this distribution was much less accurate than the log-normal distribution. As a numerical example, for one table, if $n_{ij} = 0$, and $n_{i.}n_{.j}/N = 5$, then p_{ij} was estimated as $1.87(1 \pm 0.60)$.

The insensitivity with respect to the assumed semi-initial distribution must be emphasized. Presumably a wide class of parametric distributions, with the parameters estimated from the contingency tables themselves, would have given rise to much the same numerical results. In particular, the Dirichlet distribution, which we used in the previous section, could be used, but I have not calculated its numerical implications for this problem.

7.1 Correlated Rows and Columns and Weighted Lumping

The method used was called the "association-factor" method for estimating small probabilities in a contingency table. It ignores information that is provided by correlations between the rows and between the columns of the table. The reference contains only a brief description of how this correlation information might be allowed for, under the name "weighted lumping." The idea of weighted lumping is that each row of the contingency table can be enriched by adding to it multiples of other rows that it appears to resemble. Any such technique of weighted lumping makes some allowance for the possibility that the various attributes resemble others more or less closely. This is a feature of a two-way classification (contingency table) that makes it qualitatively different from a one-way classification (multinomial distribution).

I believe that the notion of weighted lumping is more important for the estimation of small probabilities than is the main theme of the reference under discussion, so it is now discussed in more detail.

Let $\rho_1(i, i')$ and $\rho_2(j, j')$ be the sample product-moment correlation coefficients between rows i and i' and between columns j and j'. The crudest estimate for p_{ij} is n_{ij}/N, where N is the total sample size. But if n_{ij} is small, especially if it is 0 or 1, we would certainly not trust this estimate, and the methods of the reference are much better in my opinion. If however the table is not pure, and it seldom is in practice, the use of the correlations is appropriate. In a sense, the use of these correlations is implicit in actuarial practice. The usual actuarial practice is to enlarge the class or choose an analogous class; for example, the probability that a professional chess player will be kicked to death by a horse might be estimated by the probability that a man of sedentary habits will suffer from the Bortkiewitz effect. *How* to enlarge the class, or to choose the analogous classes, is liable to be a matter for judgment, but the purpose of using the correlation for weighting is to decrease the need for judgment. A possible class of estimates would be

$$\frac{\sum_{i',j'} \left\{ \frac{n_{i'j'}}{N} \cdot \frac{n_{i.}n_{.j}}{n_{i'.}n_{.j'}} f_1(\rho_1(i,i')) f_2(\rho_2(j,j')) \right\}}{\sum_{i'} f_1(\rho_1(i,i')) \sum_{j'} f_2(\rho_2(j,j'))} \tag{7.1}$$

where f_1 and f_2 are increasing functions of their arguments, with $f_1(1) = f_2(1) = 1$, and very small when the arguments are far from 1. The formula would estimate p_{ij} as a weighted average of all the ratios

$$\frac{n_{i'j'}}{N} \frac{n_{i.}n_{.j}}{n_{i'.}n_{.j'}}$$

each of which is itself an estimate of p_{ij}, when the rows (i, i') are strongly correlated, and the columns (j, j') are strongly correlated. Since the log-association-factors, or amounts of mutual information,

$$I(i, j) = \log p_{ij} - \log p_{i.} - \log p_{.j} \tag{7.2}$$

appear to have approximately normal distributions over large contingency tables, at least in the two examples in Good,[31] it seems to me to be more natural to estimate the amounts of information by means of the formula

$$I_2(i, j) = \frac{\sum_{i',j'} I_1(i', j') f_1(\rho_1(i, i')) f_2(\rho_2(j, j'))}{\sum_{i'} f_1(\rho_1(i, i')) \sum_{j'} f_2(\rho_2(j, j'))} \tag{7.3}$$

Here the first approximations $I_1(i, j)$ are obtained as in Good,[31] by the association-factor method without lumping.

Neither of the two previous formulas is open to the objection that was raised on page 114 of Good[31] to the formulas given there, that they are unreasonable when the rows and columns are perfectly correlated. Furthermore, Formula 7.3 gives a possible solution to the problem of combining a weighted-lumping method with the association-factor method, a problem that was proposed in Good,[31] p. 121, and for which no solution has yet been suggested.

The functions f_1 and f_2 should depend on the size of n_{ij}. If n_{ij} is very large, we should take $f_1(\rho) = f_2(\rho) = 0$ when $\rho \neq 1$.

7.2   Application of "Botryology" to Weighted Lumping

Let us now recall a geometrical representation of correlation coefficients. Our contingency table can be used for estimating the correlation coefficients and the partial correlation coefficients between the s attributes $A_1, A_2, \cdots, A_s$ corresponding to the rows. The entire set of correlations can be geometrically pictured by representing each attribute by a point on the $(s - 1)$-dimensional unit sphere (surface of an s-dimensional ball), in such a manner that the correlation coefficients are equal to the cosines of the lengths of the arcs of great circles joining pairs of points. The partial correlations also have geometrical significance, for example, in three dimensions they are the cosines of the angles of the spherical triangle formed by A_1, A_2, A_3. (See, for example, Kendall,[70] p. 379.) If s is large, then clusters or clumps of points on the sphere will suggest new classifications. Since there can be clusters of clusters, and so on, we see that a large contingency table can provide information for very elaborate classifications. These remarks apply of course to columns of the contingency table as well as to the rows.

It can happen that the contingency table is very sparse indeed. This is liable to happen, for example, if the rows correspond to index terms and the columns to abstracts, in problems of information retrieval. In such cases the estimates of the correlation coefficients might be very poor, and a different approach to clumping will need to be used, having from the start a less classical appearance. (See, for example, Good.[41, 43] In the former reference, the science of clumps is called "botryology," from the English combining form "botry-," or from the Greek βοτρυσ, a cluster of grapes.)

If clumps can be defined even though the correlation coefficients cannot be estimated, then estimates for the amounts of information $I(i, j)$ can be made when $n_{ij} = 0$, in the following manner. If i belongs to a clump C, and j to a clump D, then $I(i, j)$ could be estimated by $I(C, D)$, where $I(C, D)$ is the association factor between C and D. The latter can in its turn be estimated by *contracting* the original contingency table by lumping together the rows corresponding to the elements of C and lumping the columns corresponding to the elements of D. If i belongs to two overlapping clumps, then this method of estimating $I(i, j)$ is subject to ambiguity, just as the actuary's estimate of probabilities depends on his choice of reference classes. A weighted average of all the relevant $I(C, D)$'s could be used as the estimate of $I(i, j)$; more theory is required for selecting the weights. (Compare Formula 7.3.)

7.3   The Singular Decomposition of a Contingency Table

There is another completely different approach to the analysis of contingency tables. It would apply also to "population" contingency tables (see Chapter 9), but I do not know how to extend it to multi-dimensional contingency tables. In this chapter we have assumed that the null hypothesis of independence of rows and columns of the table has been rejected, and we have then turned to ideas in which the notion of independence is completely forgotten. If the hypothesis of independence had been adopted, we would have approximated the contingency table, regarded as a matrix, by means of a matrix of rank 1. A natural generalization then is to set up a sequence of null hypotheses: that the matrix (p_{ij}) is of rank 2, that it is of rank 3, and so on. Let the hypotheses that this matrix is of rank r be denoted by H_r. (The symbol r should not be confused with its meanings in other chapters.) The acceptance of any of the hypotheses H_r leads in a very natural algebraic manner to a method of estimating p_{ij}. The theory is essentially that used in factor analysis. It can best be described in terms of the *singular decomposition* of a matrix $\mathbf{A}$, where, in our application, we would take $\mathbf{A} = (n_{ij})$. We conclude this chapter with a description of the singular decomposition and some of its properties.

Let $\mathbf{A} = (a_{ij})$ be a real matrix of s rows and t columns. If there exist s-dimensional and t-dimensional vectors $\mathbf{x}$ and $\mathbf{y}$, of unit lengths, and a number κ such that

$$\left.\begin{array}{l} \mathbf{Ay} = \kappa\mathbf{x} \\ \mathbf{A'x} = \kappa\mathbf{y} \end{array}\right\} \qquad (7.4)$$

then κ is called a *singular value* of the matrix, and $\mathbf{x}$ and $\mathbf{y}$ are called *singular vectors* of $\mathbf{A}$. (The prime denotes transposition.) The terminology is basically the same as that used by Smithies[105] for kernels of integral equations. He attributes the use of singular values and singular functions of a linear transformation in Hilbert space to E. Schmidt. The theory of

singular values and singular vectors of matrices is more elementary
than the corresponding theory for kernels, but the tract by Smithies is
the most convenient reference that I have found. He uses the term
"singular value" for the reciprocal of ours.†

It will be observed that $AA'x = \kappa^2 x$, $A'Ay = \kappa^2 y$, so that κ^2 is an
eigenvalue of the Gram matrices AA' and $A'A$, x is an eigenvector of
AA', and y is an eigenvector of $A'A$. We could call x a *left singular
vector* of A and y a *right singular vector* of A. These vectors are real.

Here AA' and $A'A$ are square symmetric matrices whose eigenvalues
are the same, except that the larger matrix of the two (when A is not
square) has a number of zero eigenvalues extra. Both matrices can be
shown to be non-negative definite, so that all the eigenvalues are
non-negative. They therefore have real square roots, and the positive
values of these square roots are taken here, by convention, as the singular
values of A. We think of them as ordered in decreasing order,
$\kappa_1 \geq \kappa_2 \geq \cdots \geq \kappa_\mu$, $\mu = \min(s, t)$, and we then refer to κ_r as the rth
singular value of A. For the sake of simplicity, we shall assume that
the singular values are all distinct. Then the corresponding left and
right singular vectors are unique if they are normalized to have unit
length. We denote them by x^r and y^r, which should not be read as
powers. The x^r's form an orthonormal set, and so do the y^r's.

It is familiar (see, for example, Bellman,[4] p. 94) that a symmetric
matrix can be diagonalized by means of an orthogonal transformation,
where the orthogonal matrix has columns that form a complete set of
eigenvectors of the symmetric matrix. In other words, a symmetric
matrix can be decomposed into a sum of matrices each of rank 1, thus:
$B = \sum \lambda_r x^r(x^r)'$, where the λ's are the eigenvalues of B (which are real),
and the x^r's form an orthonormal set of eigenvectors. Equivalently, a
quadratic form can be transformed into the form $\sum \lambda_r x_r^2$ by means of
an orthogonal transformation. The spectral decomposition of a self-
adjoint linear transformation of a finite-dimensional vector space
(Halmos,[49] p. 156) is again another way of saying the same thing.

The following decomposition of an arbitrary real matrix is not easy to
find explicitly in books on matrices, although the more difficult case of the
decomposition of a linear transformation in Hilbert space is given, for
example, by Smithies,[105] p. 147 (the analogue of "Hilbert's formula").

The decomposition of a matrix, which we call the *singular decomposition*, is

$$A = \sum_{r=1}^{\mu} \kappa_r x^r(y^r)' \qquad (\mu = \min(s, t)) \qquad (7.5)$$

$$a_{ij} = \sum_r \kappa_r x_i^r y_j^r \qquad (7.6)$$

† I have now found the further reference, Whittle.[111]

Equivalently, there exist two orthogonal matrices, $\mathbf{X}$ ($s \times s$) and $\mathbf{Y}$ ($t \times t$) such that

$$\mathbf{XAY} = \begin{bmatrix} \kappa_1 & 0 & & 0 & 0 & 0 & \cdots & 0 \\ 0 & \kappa_2 & & 0 & 0 & 0 & \cdots & 0 \\ & & \ddots & \vdots & \vdots & \vdots & & \vdots \\ 0 & \cdots & & \kappa_s & 0 & 0 & \cdots & 0 \end{bmatrix} \text{(if } s < t\text{)} \quad (7.7)$$

We see at once that, for any integer ν ($\nu = 0, \pm 1, \pm 2, \cdots$)

$$(\mathbf{AA}')^\nu = \sum_r \kappa_r^{2\nu} \mathbf{x}^r (\mathbf{x}^r)' \quad (7.8)$$

$$(\mathbf{A}'\mathbf{A})^\nu = \sum_r \kappa_r^{2\nu} \mathbf{y}^r (\mathbf{y}^r)' \quad (7.9)$$

$$(\mathbf{AA}')^\nu \mathbf{A} = \sum_r \kappa_r^{2\nu+1} \mathbf{x}^r (\mathbf{y}^r)' \quad (7.10)$$

On putting $\nu = 0$ we get Equation 7.5 and the decompositions of the two identities:

$$\mathbf{I}_s = \sum_{r=1}^{\mu} \mathbf{x}^r (\mathbf{x}^r)' \quad (7.11)$$

$$\mathbf{I}_t = \sum_{r=1}^{\mu} \mathbf{y}^r (\mathbf{y}^r)' \quad (7.12)$$

There is a natural analogue of the familiar "power method" for finding the eigenvalues and eigenvectors of an arbitrary matrix, as given, for example, by Duncan, Frazer, and Collar,[17] p. 140. We start with an arbitrary vector and successively multiply it by $\mathbf{A}$ and $\mathbf{A}'$, with normalization of the length of the vectors obtained on the way. This process will converge with probability 1, as can be readily seen from the singular decomposition of the matrix.

Once the leading singular value and singular vectors have been computed, the first term of the singular decomposition is known and can be subtracted from $\mathbf{A}$. Then one can continue to compute the next singular value κ_2 and the left and right singular vectors that belong to it, and so on.

7.4 A Sequence of Hypotheses for the Cell Probabilities

One of the uses of the singular decomposition of a kernel is that it leads to a sequence of approximations to the kernel, of finite rank (Smithies,[105] p. 148). Similarly the singular decomposition of a matrix provides a finite sequence of approximations to the matrix, of ranks 1, 2, 3, $\cdots$. Basically, this is the idea behind factor analysis (see, for example, Harman,[52] p. 13), to approximate a matrix, whose rows have been made to add up to zero, by means of a matrix of as low a rank as possible. The approximation of rank μ is of course exactly equal to the original matrix A. For kernels, the approximation is in mean square, and the same is true for the approximations to a matrix. In fact, it is straightforward to see that the minimum value of

$$\sum_{i,j} (a_{ij} - \alpha x_i y_j)^2 \tag{7.13}$$

where α is real and the vectors x and y are real and of unit lengths, is attained when α is the leading singular value κ_1, and x and y are the corresponding singular vectors. The minimum value is $\kappa_2^2 + \kappa_3^2 + \cdots + \kappa_\mu^2$. By then defining a new matrix $B = A - \kappa_1 xy'$, we see that the minimum of

$$\sum_{i,j} (a_{ij} - \alpha_1 x_{1i} y_{1j} - \alpha_2 x_{2i} y_{2j})^2 \tag{7.14}$$

is attained when $\alpha_1 = \kappa_1$, $\alpha_2 = \kappa_2$, and the x's and y's are the first two pairs of singular vectors. The minimum value is $\kappa_3^2 + \cdots + \kappa_\mu^2$. And so on.

In order to test the hypothesis H_r that the matrix of population probabilities is of rank r, we could put

$$p_{ij}^{(r)} = \kappa_1 x_{1i} y_{1j} + \kappa_2 x_{2i} y_{2j} + \cdots + \kappa_r x_{ri} y_{rj}$$

and then apply the chi-squared test, with

$$\chi^2 = \sum_{i,j} \frac{(n_{ij} - N p_{ij}^{(r)})^2}{N p_{ij}^{(r)}} \tag{7.15}$$

which has $(s - r)(t - r)$ degrees of freedom. In order to see that this is the number of degrees of freedom, note that, for each ρ ($\rho = 1, 2, \cdots, r$), the vector $(x_{\rho i})$ ($i = 1, 2, \cdots, s$) is of unit length and is orthogonal to all the $\rho - 1$ earlier singular vectors (x_{1i}), (x_{2i}), $\cdots$, $(x_{\rho-1, i})$. It is therefore determined by $s - \rho$ parameters. So the number of parameters that are varied when determining $p_{ij}^{(r)}$ is

$$(s + t - 1) + (s + t - 3) + \cdots + (s + t - 2r + 1)$$
$$= st - (s - r)(t - r)$$

For factor analysis in general, one does not start with an array of frequencies, and the chi-squared test is therefore not available. The usual test is due to D. N. Lawley and M. S. Bartlett. (See, for example, Harman,[52] p. 371.) I do not know which test is better when we do start with an array of frequencies. There are connections between singular values of a square matrix and its eigenvalues. An example of such a connection was given by Weyl.[109] He proved, for example, the following result:

*Let $h(\lambda)$ be an increasing function of its positive argument, $h(\lambda)$
$\geq h(\lambda')$ for $\lambda \geq \lambda' > 0$, such that $h(e^\xi)$ is a convex function of ξ and
$h(0) = \lim_{\lambda \to 0} h(\lambda) = 0$. Then the eigenvalues λ_r and the singular values κ_r
in descending order† satisfy the inequalities*

$$\phi(\lambda_1) + \cdots + \phi(\lambda_r) \leq \phi(\kappa_1) + \cdots + \phi(\kappa_r) \qquad (r = 1, 2, \cdots, \mu)$$

In particular

$$\lambda_1^\alpha + \cdots + \lambda_r^\alpha \leq \kappa_1^\alpha + \cdots + \kappa_r^\alpha \qquad (r = 1, 2, \cdots, \mu)$$

for any real exponent $\alpha > 0$.

Weyl mentions that his results "carry over completely to completely
continuous linear operators in Hilbert space, especially to continuous
kernels of integral equations." The same remark applies to the present
discussions, and to those in Appendix D.

It is natural to ask whether the singular decomposition of a continuous
kernel would have useful statistical applications by analogy with the
discrete case, which is factor analysis. In the application of factor
analysis to the rectangle of marks obtained by several people in several
examinations, no natural ordering is assumed for the various people
nor for the various examinations. Suppose however we were interested
in the yield per acre of some farm produce, when two distinct treatments
were simultaneously applied to various extents, say x and y. For each
pair (x, y) of extents of the two treatments, there would be an observed
yield, $f(x, y)$ say, and we could now consider the singular decomposition
of $f(x, y)$ regarded as a continuous kernel. We would of course not have
observed values of $f(x, y)$ for all values of x and y, but we could make
estimates by means of some method of surface fitting. An approximation
to $f(x, y)$, of low rank, would have some possible physical significance.
For example, if $f(x, y)$ factorized in the form $f(x)g(y)$, then the two
treatments would be statistically independent and their effects would
be multiplicative. If this suggestion for extending the method of factor
analysis is used, one would consider transforming the coordinates x and
y, in order that $f(x, y)$ should, in some sense, approximately factorize.

A hybrid suggestion can also be made, in which one treatment is
discrete and unordered, and the other is ordered and continuous. Then
we should have a cross between a matrix and a kernel, $f_i(y)$ say; in
other words, we should have a finite sequence of continuous functions. In
this case too there is no difficulty in defining singular values and a
singular decomposition, and a truncation of the singular decomposition

† Note: Readers of Weyl's paper are warned that his λ's and κ's are the squares
of ours.

would constitute an approximation to the "kertrix" or whatever term is appropriate for the hybrid.

In Appendix D, a minimax property of singular values is discussed, partly for its mathematical interest and partly because it is liable to have some practical application.

8. The Sampling of Species[†]

Suppose that a random sample is drawn from an infinite population of animals of various species, or of words from a literary text. Let the sample size be N and let n_r distinct species be each represented exactly r times in the sample, so that

$$\sum r n_r = N \qquad (8.1)$$

The number n_r is the frequency of the frequency r. The sample tells us the values of $n_1, n_2, \cdots$, but not usually of n_0. In Good[29] and Good and Toulmin,[44] suggestions were made for estimating, among other things,[‡]

1. The population frequency of each species.
2. The total population frequency of all species represented in the sample, or, as we may say, the *coverage* of the sample.
3. Various population parameters measuring heterogeneity, including "entropy."
4. The number of distinct species and the coverage of a second sample of size λN, where $\lambda < 4$.

The results are applicable, for example, to samples of vocabulary by makers of dictionaries, of accident proneness, occurrences of chess openings, and species of plants or animals.[*] An example is Eldridge's

[†] The notation in this chapter is not usually the same as in previous chapters.
[‡] See also B. Harris.[53]
[*] Dr. Alan B. Shaw has recently pointed out (private communication) an important application to industrial paleontology.

sample in 1911 of 43,989 fully inflected words of American newspaper English, for which the number of *distinct* words (species) that occurred was 6001. Here $n_1 = 2976$, $n_2 = 1079$, $n_3 = 516$, $n_4 = 294, \cdots, n_{4290} = 1$. (Quoted in Good,[29] p. 257, from Zipf,[115] pp. 64 and 25.)

In what manner does this sampling-species problem differ from multinomial sampling, which was considered in Chapters 4 and 5? The answer is that extra information can be derived from the frequencies of the frequencies, provided that they can be smoothed. Some smoothing is possible, at least for small values of r, when n_1 is large. *Johnson's sufficiency postulate certainly ignores information when such smoothing can be done with any confidence.* Consequently the appropriate theory for the species-sampling problem does not make use of the Type II Dirichlet distribution.

If a particular species is represented r times in the sample, then the maximum-likelihood estimate of its population frequency is r/N. If this estimate were used in order to estimate the coverage of the sample, we would estimate the coverage as 100 per cent, which is completely unreasonable.

The theory in the two references began from a suggestion made by A. M. Turing (private communication) in 1941. He proposed that the probability should be estimated, not by r/N, but by r^*/N, where r^* is an adjusted value of r given by the formula

$$r^* = \frac{(r + 1)n_{r+1}}{n_r} \tag{8.2}$$

Herbert Robbins has recently pointed out the close resemblance between this formula and formula (19) of Robbins,[96] and between the argument given in Good[29] for justifying Equation 8.2 (or rather Equation 8.3) and the empirical Bayes method described by him. (See also Robbins.[97])

Really one should replace Equation 8.2 by

$$r^* = \frac{(r + 1)n'_{r+1}}{n'_r} \tag{8.3}$$

where (n'_r) is a smoothing of (n_r). A fair amount of space is allotted in Good[29] to methods of smoothing.

We shall not here take space to give the proof of Equation 8.3, since it is given in detail, together with many other results, in the references. The reader who does not wish to look up the references will find it instructive to consider the special case when n'_r is of the Poisson form $se^{-a}a^r/r!$ Then r^* reduces to a constant, independent of r, *as it should.*

From Equation 8.2 or 8.3, we can at once deduce the estimate $1 - n_1/N$ for the coverage of the sample; in other words, the probability that the next animal sampled will belong to a new species is (approximately) equal to the proportion of animals in the existing sample belonging to species that were represented only once. An estimate for the word-by-word entropy, from a sample of vocabulary, is

$$\log N - \frac{1}{N} \sum_r r n_r' \left(1 + \frac{1}{2} + \cdots + \frac{1}{r} - \gamma + \frac{d}{dr} \log n_r' \right) \tag{8.4}$$

from which an estimate of letter-by-letter entropy could be deduced by treating words of each length separately. The expected number of new species in an independent sample of size λN is approximately

$$\lambda n_1 - \lambda^2 n_2 + \lambda^3 n_3 - \cdots \tag{8.5}$$

if $\lambda < 1$. If $1 < \lambda < 4$, this formula can be transformed and made usable by certain methods of transforming divergent series, such as Euler's method (Hardy[51]) or Shanks' method (Shanks[101]).

An interesting method of smoothing, not mentioned explicitly in the references, is based on such methods of transformation of series. Let n_r ambiguously denote not merely the frequency of the frequency r but also the corresponding random variable, for a sample of size N, and let $n_r(2)$ denote the corresponding random variable for a sample of size $2N$. Then, in accordance with equation (14) of Good and Toulmin,[44] we have

$$\mathscr{E}(n_r(2)) = \sum_{i=0}^{N-r} \frac{\mathscr{E}(n_{i+r}) \binom{2N}{r} \binom{-N}{i}}{\binom{N}{i+r}} \tag{8.6}$$

and

$$\mathscr{E}(n_r) = \sum_{i=0}^{N} \frac{\mathscr{E}(n_{i+r}(2)) \binom{N}{r} \binom{N}{i}}{\binom{2N}{i+r}} \tag{8.7}$$

The series in Equation 8.6 has alternating signs. If we replace $\mathscr{E}(n_{i+r})$ by the observed value of n_{i+r}, and apply a transformation of the series and truncate it, we obtain a smoothing of $\mathscr{E}(n_r(2))$. By substituting in Equation 8.7, we obtain a smoothing of (n_r). We can then use these smoothed values again in Equation 8.6, and so on iteratively. This method should be better than the method suggested on page 245 of Good,[29] since that method was based on less accurate formulas. Any method of smoothing should however be tested by the chi-squared test, as suggested on the same page of the latter reference. The Equations 8.6 and 8.7 cannot be used iteratively, without intermediate smoothing, since they imply one another.

Sometimes, but not always, the frequencies seem to obey the so-called Zipf law† for the distribution of words in a language (Zipf,[115] Mandelbrot,[82] Simon,[103] Good,[32] and Herdan,[55] or the Pareto law for the distribution of incomes (Champernowne[10]), in which case these "laws" can be used for smoothing and the present theory applied. But for random (multinomial) sampling, the Zipf law should not be taken with very great seriousness, at any rate for rare words, since this theory shows that this law is not invariant, whatever its parameter, when the sample size is increased. This was pointed out in Good,[37] where it was also pointed out that Simon's explanation of the Zipf law is certainly inconsistent with the assumption of multinomial sampling. Mandelbrot's explanation does apply to multinomial sampling, hence both explanations could be correct, since they apply to distinct problems.

Recently, a new law has been suggested by Belonogov,[5] for Russian vocabulary, which seems to be more accurate than Zipf's law, but it contains more parameters. The author was apparently unaware of Zipf's law.

The methods of this chapter ignore any information available from the knowledge that some species strongly resemble each other. It is difficult to formulate a general theory that would enable this additional information to be taken into account, and, even if there were such a theory, it would be liable to be complicated to use. If it were used, the sampling problem could hardly be regarded as a problem in multinomial sampling alone, since the additional information would not be incorporated in the frequency count, and not even in the frequencies of the frequencies.

Thus we see that the apparently simple problem of estimating the probabilities of words, in a given population of text, is not very simple. For the design of an ultraintelligent machine capable of information retrieval, one would like to estimate the probabilities of references, in texts, to assigned objects and concepts. This would be more complicated partly because of the difficulty of deciding what is meant by an "object." A botryological suggestion for resolving this problem has been made by K. Spärck Jones.[68]

In spite of the complications in a complete statistical analysis, the human mind seems to have some facility in estimating the probabilities of words, down to probabilities of one in a million, without an explicit

---

† A better form of Zipf's law is expressed in terms of the population frequencies, and was explained by Mandelbrot in terms of maximization of amount of information per unit of effort. In my brief exposition[32] of his explanation there were two minor errors: one was "minimize" in place of "maximize." It included an explanation of the distribution of the number of strokes in symbols in Pitman's shorthand.

sample. With a little practice based on word-frequency lists, such as Thorndike and Lorge,[107] one can usually estimate the probabilities within a factor of 10. An ultraintelligent machine, must by its definition, be capable of a better performance.

We have perhaps said enough now to indicate the interest of the species-sampling problem. It is a very good example of the estimation of probabilities from an effectively small sample, when the sample size is in fact large.

9. Multidimensional Population Contingency Tables

We have so far considered the estimation of probabilities for one-way classifications (multinomial distributions, including the sampling of species) and two-way classifications (two-dimensional contingency tables). Multidimensional classifications occur, for example, in pattern recognition, in information retrieval, and in human recall: a document might be classified by topic, author, size, color, and language, or a memory might be recalled by asking different kinds of questions, such as when?, where?, why?, with what?, and to whom? (See, for example, Maxwell.[84])

There is some known theory concerning the estimation of probabilities in multidimensional classifications. Much of it can be expressed in terms of *multidimensional contingency tables*. Such a table is a multi-dimensional array (p_i) of physical probabilities, where i is a vector of integers,

$$i = (i_1, i_2, \cdots, i_m)$$

$$0 \le i_1 \le d_1 - 1, \cdots, 0 \le i_m \le d_m - 1$$

Here (p_i) is an m-dimensional population contingency table, specifically a $d_1 \times d_2 \times \cdots \times d_m$ table.

Suppose that we do not know all the p_i's, but we know some relationships between them (*constraints*) such as some of the marginal totals. This can happen because the sample from which the table is inferred is not large enough to estimate the p_i's individually and directly. Indeed the corresponding frequencies n_i might all be 0 or 1. How can we make estimates of all the p_i's? There is nothing in the axioms of probability that can help us; but a technique that seems largely justifiable, judging by its consequences, is to maximize the entropy, $-\sum p_i \log p_i$, subject to the constraints.

The principle of maximum entropy was first used by Boltzmann[6] and Gibbs[26] for application to statistical mechanics. They found that the principle implied the whole of classical equilibrium thermodynamics. Shannon[102] mentions that if the principle is applied to a real random variable of given variance, then the variable must be Gaussian, and he generalizes this to multidimensional variables. Jaynes[61, 62] applied the principle to the vexed problem of selecting credibilities, in order to generalize the Bayes-Laplace postulate of a uniform distribution. For example, if $p_1 + p_2 + \cdots + p_6 = 1$, then the distribution of maximum entropy is $p_1 = p_2 = \cdots = p_6 = 1/6$, and these are suggested as the rational betting probabilities (credibilities). The present writer[42] proposed a somewhat different interpretation:

Let X be a random variable whose distribution is subject to some set of constraints. Then entertain the null hypothesis that the distribution is the one of maximum entropy subject to these constraints.

With this interpretation, the principle is, for example, to be regarded as generating classical statistical mechanics as a null hypothesis, to be tested experimentally. Jaynes' interpretation is philosophically different. The philosophical point of view makes a practical difference in the application of the principle of maximum entropy to multidimensional contingency tables. (In this application the principle can be expressed by saying that the expected amount of mutual information between rows, columns, spurs, and so forth, is minimized: see Hartmanis,[54] P. M. Lewis II,[79] D. T. Brown.[8])

For a two-dimensional population contingency table with assigned marginal totals, the principle leads to the null hypothesis of no association. A similar conclusion applies to an m-dimensional table when the only marginal totals are those corresponding to complete $(m-1)$-dimensional blocks. In general we could describe the effect of the principle of maximum entropy by saying that, in some sense, it pulls out the hypothesis in which the amount of independence is as large as possible.

It would be interesting to know whether the hypotheses H_r ($r = 2, 3, \cdots$) of Section 7.4 correspond to local maxima of the constrained entropy, and if so whether there is a generalization to more than two dimensions.

For a 2^m table, with *all* marginal totals assigned, the principle leads to the equation

$$\prod_{i}^{|i|\,\text{even}} p_i = \prod_{i}^{|i|\,\text{odd}} p_i \tag{9.1}$$

where $|i|$ is the number of nonzero components of i. This equation determines all the p_i's uniquely. For the case $m = 3$, it turns out that the principle leads to the same determination of the p_i's as suggested by Bartlett[3] many years ago, for different reasons. He described the condition for this case as that of the vanishing of the second-order interactions. For a three-dimensional table, independence is equivalent to the vanishing of the first-order and second-order interactions, as described next.

If we know the sums of p_i over each subset of $m - r$ coordinates, we say that we have *a complete set of* rth-order constraints ($r = 0. 1, \cdots, m - 1$). It turns out that, when we have a complete set of rth-order constraints in an m-dimensional 2^m table, the principle of maximum entropy leads to the null hypothesis that the rth-order and all higher-order interactions vanish; where, by definition, the m-dimensional mod 2 discrete Fourier transform,

$$I_j = \sum_i (-1)^{i_1 j_1 + \cdots + i_m j_m} \log (2^m p_i) \tag{9.2}$$

is called an interaction, or a Fourier log-interaction of order $|j| - 1$. It is analogous to an interaction in a factorial experiment (see, for example, Good.[35]) A similar conclusion applies for the general m-dimensional population contingency table, with an appropriate definition for interaction. The generalized definition of interaction can be expressed in several equivalent manners. It will suffice here to say that the interactions of order at least s all vanish if and only if they all vanish in each 2^m subtable. The reader is referred to Reference 35 for a complete statement and proof, and for a discussion of related matters, such as significance tests for the hypothesis of no rth-order or higher-order interaction within the wider hypothesis of no sth-order or higher-order interaction, when we have a *sample* table (n_i). (See also, for example, Darroch,[14] and Goodman,[45] both of whom give further references.) There is also a discussion of interactions in Markov chains.

We quote one more result here, giving a curious duality connecting maximum-likelihood with maximum entropy:

Suppose that we have a sample (n_i), of total size $n = \sum_i n_i$. Then the maximum-likelihood estimates of the p_i, subject to the vanishing of the *rth and all higher-order interactions, are equal to the maximum-entropy values of the* p_i*, subject to the rth-order sums of the* np_i*'s being equal to those of the* n_i*'s provided that the maximum-likelihood is reached at a stationary value of the likelihood.*

As a consequence of this result, the maximum-likelihood estimates of the p_i's could be obtained numerically by means of the *iterative scaling procedure* for maximizing entropy (D. T. Brown[8] and Darroch[14]). In this procedure the constraints are taken in order cyclically and the p_i's involved in a constraint are scaled so as to satisfy the constraint.

Let us now return to the question of philosophical interpretation. Let p be a probability derived from the principle of maximum entropy. In Jaynes' interpretation, p would be a rational betting probability, whereas, for us, it would be a rational betting probability only if we accept the null hypothesis. Since, subjectively, the null hypothesis is uncertain, we cannot regard p as a betting probability with much confidence. On the other hand, if we have a null hypothesis and no rival one, it sometimes seems fairly reasonable to accept this hypothesis provisionally. In these circumstances Jaynes' interpretation and ours should lead to similar decisions.

Another method of estimating (p_i), when we are given a multidimensional contingency table (n_i), instead of finding null hypotheses that cannot be rejected, is to use the least-squares method of Deming and Stephan,[16] to which we referred in Chapter 7. As in the two-dimensional case, their methods are convenient and useful but suffer from the same disadvantage of estimating p_i as zero when $n_i = 0$. When the dimensionality is high, this is a very serious disadvantage. Perhaps this disadvantage could be overcome by minimizing

$$\chi^2 = \sum \left\{ \frac{(p_i - n_i/N)^2}{Np_i} \right\}$$

subject to the constraints, instead of minimizing

$$\sum \frac{(p_i - n_i/N)^2}{n_i}$$

Unfortunately the method of minimum χ^2 leads to heavy calculations, and I have tried no examples of it.

Another possible approach is mentioned in Good,[42] p. 931, namely the maximization of a linear combination of the entropy and the log-likelihood. This method has some aesthetic appeal but is difficult to justify. It is equivalent to the selection of the distribution of maximum final credibility, assuming the logarithm of the initial density to be proportional to the entropy. This is of course not consistent with the use of a Dirichlet distribution; but all methods so far suggested have a degree of arbitrariness. Since the initial (Type II) density is of the form

$$\prod p_i^{-kp_i} \tag{9.3}$$

it is tempting, by analogy with the methods of Chapters 5 and 6, to assume a Type III distribution for k, in other words to assume a Type II initial density of the form

$$\int_{-\infty}^{\infty} \frac{\prod p_i^{-kp_i} \, dF(k)}{\operatorname*{aver}_{\mathbf{p}} \prod p_i^{-kp_i}} \tag{9.4}$$

If the calculations could be done, this approach might very well lead to a reasonable solution for most multidimensional contingency tables.

The only calculations I have made, using this kind of initial distribution, applies to binomial sampling. The initial density proportional to $p^{-p}(1 - p)^{-(1-p)}$ was assumed, and the estimates (expected values) of p were computed for r successes out of N trials, for small values of N. The results are given in Column iii of Table 9.1. The Bayes-Laplace

TABLE 9.1

r	N	i	ii	iii
0	0	0.50	0.50	0.50
0	1	0.33	0.25	0.36
0	2	0.25	0.17	0.28
1	2	0.50	0.50	0.50
0	3	0.20	0.13	0.23
1	3	0.40	0.38	0.41
0	4	0.17	0.10	0.19
1	4	0.33	0.27	0.35
2	4	0.50	0.50	0.50
0	5	0.14	0.08	0.17
1	5	0.29	0.25	0.31
2	5	0.43	0.42	0.44

estimate is given in Column i and the Perks-Jeffreys estimate in Column ii for comparison. At least in this example, the Bayes postulate leads to much the same results as the assumption of a density proportional to the antilogarithm of the entropy, so presumably Formula 9.4 would lead to much the same conclusions as the weighted sum of symmetrical Dirichlet distributions.

10. Summary

If, in a binomial sample of size N (N trials) there are r *successes*, then Laplace's estimate of the *physical probability* p of a success is $(r + 1)/(N + 2)$. This estimate is the subjective or credibilistic or Type II expectation of p if the initial (prior) Type II distribution of p is uniform (the Bayes *postulate*). The epithet *Type II* is used here as a neutral expression, since the initial distribution *could* be a physical probability distribution. (Kinds of probability are briefly discussed in Chapter 2.) This formula is usually called *Laplace's law of succession*, since it is supposed to represent the Type II probability that the very next trial will be a success. Its square is not supposed to be the probability that the next two trials will be successful!

In Chapter 3, this problem of binomial sampling (more specifically *simple sampling*) is discussed in detail. The precise relationship between binomial and simple sampling was first established by de Finetti, but some of the ideas had previously been published by W. E. Johnson and Julius Haag.

In Chapter 4, the discussion is extended to multiple sampling (t categories, where $t > 2$), and here W. E. Johnson's ideas are especially relevant. In effect he advocated, in certain circumstances, an estimate for multinomial probabilities, which we describe as the use of a *flattening constant k*. In Laplace's law of succession this constant is 1. A short proof of the appropriate extension of de Finetti's representation theorem is given, and, from the uniqueness of this representation, it follows that the use of a flattening constant k is equivalent to the use of an initial

Type II distribution of the symmetrical Dirichlet form. Johnson's argument depends especially on a postulate, here called his *sufficiency postulate*, that cannot be doubted when $t = 2$; but unfortunately his deduction of the validity of a flattening constant breaks down in this case, a fact that seems to have been generally overlooked. When $t > 2$, Johnson's sufficiency postulate seems intuitively a reasonable approximation for small samples, for which it is mainly required, but he gave no rules for determining k. The value $k = 0$ is equivalent to maximum-likelihood estimation and contradicts the Bayesian philosophy. It seems reasonable in some circumstances to assume that the initial Type II distribution is a weighted sum of all symmetrical Dirichlet distributions, that is, to use a *Type III* distribution for k. In physical terms, the symmetrical Dirichlet distribution could be regarded as selected from a super-superpopulation, or population of Type III.

At the end of Chapter 4, some problems are discussed concerned with arboresque (treelike) classifications. The idea is taken up again in Appendix E where it is found that the problem of finding a self-consistent initial distribution for the four physical probabilities of a 2×2 contingency table, can be expressed in terms of a nonlinear integral equation if Johnson's sufficiency postulate is used indiscriminately. The solution of this equation leads to a *reductio ad absurdum* and shows that the conditional physical probabilities in one row of a contingency table are relevant to the initial Type II distribution of the probabilities in any other row.

The proof of de Finetti's theorem and its extension to multiple sampling depend on the use of known facts concerning the Hausdorff moment problem, and in Appendix B it is pointed out that Hausdorff's method of summation can be regarded as a Monte Carlo process.

In Chapter 5, the ideas of Chapter 4 are applied to questions of multinomial discrimination and significance, and these questions lead (Chapter 6) to some new Bayesian tests of independence (lack of association) in contingency tables. For tables with only two rows or two columns these tests are satisfactory, but for tables with several rows and also several columns the calculations are liable to be impracticable. The practicability of the calculations however needs further study. In order to show that we are not dogmatic, an example of a large "folded" contingency table, which occurred in connection with molecular biology, is treated by an *ad hoc* method, that is, not by the above-mentioned method and not by any standard method. The reason for the use of an *ad hoc* method is that the standard methods break down, and the above-mentioned Bayesian method requires too much calculation.

In Chapter 7, we consider the estimation of small probabilities in large pure contingency tables. We include a discussion of the *singular decomposition* of a matrix, which seems to be the heart of the mathematics behind factor analysis. A significance test is given for any hypothesis concerning the rank of the underlying probability matrix. In Appendix D a minimax property of the singular values is proved. It leads to an intertwining relationship between the singular values of a contingency table and those of the table with a row deleted.

In Chapter 8, we discuss multinomial sampling when the number of categories is very large, possibly infinite (the *sampling of species*). This differs from ordinary multinomial sampling in that information is conveyed by the *frequencies of the frequencies*. The work has application to taxonomy and to the compilation of dictionaries.

In Chapter 9, multidimensional population contingency tables are considered, with special emphasis on the method of maximum entropy.

When we wish to select an initial probability distribution, all methods are fair for aiding the judgment. One method is to use some non-Bayesian approach with which one has some sympathy and see what initial distribution is implied. This can sometimes be done if Fisher's fiducial argument is used. (Fisher latterly attributed the fiducial argument to Bayes, but it is certainly "non-Bayesian" in the usual sense of the term.) But it is shown in Appendix A that the fiducial argument is not consistent with a Bayesian philosophy.

In this monograph we have used a variety of methods, mostly "Bayesian" in a modern idiom apparently never envisaged by Bayes himself. These methods are all much concerned with the foundations of probability and statistical inference and should bridge the gap between the philosopher and the practical statistician. But an infinity of questions remains unanswered in this fundamental problem of scientific inference, the estimation of probabilities.

Appendix A.   The Incompatibility of Fiducial and Bayesian Inference[†]

In this appendix it will be argued that fiducial inference is incompatible with most current theories of subjective, logical, intuitive, or inductive probability. The notation in the appendix will be self-contained.

Let us suppose that we perform two separate experiments in order to obtain information concerning a parameter θ. The results of the two experiments are real random variables x and y, having the probability density functions

$$f_1(x \mid \theta) = \frac{\theta^2(x + a)e^{-x\theta}}{a\theta + 1} \qquad (a > 0, \theta > 0, x \geq 0)$$

$$f_2(y \mid \theta) = \frac{\theta^2(y + b)e^{-y\theta}}{b\theta + 1} \qquad (b > 0, \theta > 0, y \geq 0; b \neq a)$$

This setup differs from an example of Lindley's[81] mainly in that I have taken $a \neq b$. My reason for taking $a \neq b$ will shortly be explained.

After the first experiment is performed, the fiducial probability density for θ is

$$\phi_x(\theta) = \left| \frac{\partial F_1(x \mid \theta)}{\partial \theta} \right|$$

† See Section 2.2.

81

where

$$F_1(x \mid \theta) = \int_0^x f_1(\xi \mid \theta)\, d\xi$$

$$= 1 - e^{-x\theta}\left(1 + \frac{x\theta}{a\theta + 1}\right)$$

Thus

$$\phi_x(\theta) = \frac{\theta x e^{-x\theta}}{(a\theta + 1)^2}\left[a + (a + x)(1 + a\theta)\right]$$

In accordance with Fisher,[23] p. 125, this fiducial distribution can be used as an initial (or intermediate) distribution for the next experiment. It was argued by G. A. Barnard (private communication, 1960) that this step would be illegitimate with $a = b$, for in this case the two experiments could be regarded as combined into a single random sample of the same random variable. In this case the first experiment could hardly be regarded as a "recognizable subset" of the whole experiment, and therefore it would not be in the spirit of fiducial inference to use the first experiment to set up a fiducial distribution. It would be more in the spirit of the method to use the entire evidence for this purpose. This would be legitimate since the evidence as a whole would have no "recognizable subset." (See Fisher,[22] p. 55.) This argument of Barnard's seems reasonable to me, as an interpretation of Fisher's intentions, although what Fisher states is the converse, that is, that the simplest form of the fiducial method is applicable when there are no recognizable subsets. Be this as it may, we may evade Barnard's criticism by taking $a \neq b$.

If we now use $\phi_x(\theta)$ as the initial or intermediate probability density and combine it with the result of the second experiment, we arrive at the final probability density

$$\psi_{xy} = \frac{x\theta e^{-x\theta}}{(a\theta + 1)^2}\left[a + (a + x)(1 + a\theta)\right]\frac{\theta^2}{b\theta + 1}(y + b)e^{-y\theta}$$

If the experiments are taken in the reverse order, we should arrive at a final probability density ψ_{yx}, which is obtainable from ψ_{xy} by interchanging a with b and x with y. It is almost certain that ψ_{yx} will not be equal to ψ_{xy}.

But in most current theories of intuitive probability, the final probabilities cannot depend on the order in which the two experiments happen to be reported. The axioms of these theories do not involve the times at which propositions are reported. Whichever of the two experiments was performed first, their results could be reported either way round to the mathematical statistician. It therefore follows that the fiducial method, as described by Fisher (Fisher,[22] pp. 114 and 125), where the fiducial distribution obtained from one experiment is used as the initial or intermediate distribution for a second experiment, is definitely incompatible with these theories of probability.

A method of using fiducial inference that might perhaps overcome this difficulty would be to *average the fiducial distributions over all methods of permuting the observations*. This "permutative fiducial inference," as it might be called, would sometimes be arithmetically very laborious. In the earlier example, it would however amount merely to an averaging of two distributions. But in all honesty I doubt if the fiducial argument is worth salvaging; history will presumably tell.

Appendix B.   Hausdorff Summation as a Monte Carlo Process[†]

Let s_0, s_1, s_2, $\cdots$ be the sequence of partial sums of a convergent or divergent series. Every "regular Hausdorff transformation," as defined by Hardy,[†] can be obtained in the following manner. Suppose we have a simple-sampling process for which the permutation postulate is true. To each positive integer N we associate s_r, where r is the number of successes in the first N trials. Denote the (Type II) expected value of s_r by t_N. Then the sequence (t_N) is a regular Hausdorff transform of (s_N). When t_N tends to a limit, this limit is a Hausdorff sum of the original infinite series. This property of a Hausdorff sum shows that it can be approximated by a *Monte Carlo* process.

The generalization of de Finetti's theorem to multiple sampling at once suggests a generalization of Hausdorff's summation method, to multiple series. Suppose we have a $(t-1)$-dimensional series whose partial sums are $s_\mathbf{n}$, where $\mathbf{n}$ is the vector $(n_1, \cdots, n_{t-1})$. We also have a multiple sampling process for which the permutation postulate is true. To each positive integer N we associate the partial sum $s_\mathbf{n}$, where $\mathbf{n}$ is the frequency count of the letters $1, 2, \cdots, t-1$ in the first N trials. Denote the (Type II) expected value of $s_\mathbf{n}$ by t_N. If this tends to a limit as N tends to infinity, we would naturally call this limit a generalized Hausdorff sum of the multiple series. The case $t = 2$ corresponds to ordinary Hausdorff summation.

Both ordinary Hausdorff summation and this generalization can thus be regarded as forms of a Monte Carlo summation of series.

† See Chapter 3, this volume, and Hardy,[51] Chap. 11, or Widder,[112] Chap. 3.

Appendix C.   Folded Contingency Tables[†]

Let (n_{ij}) $(i, j = 1, 2, \cdots, t)$ be a square contingency table. If the probabilities of the cells (i, j) and (j, i) are known to be equal, then it is natural to analyze the table by folding it across the main diagonal. Let $n_i = n_{i.} + n_{.i}$. We can think of n_i as a marginal total of the folded contingency table, but it should be held in mind that it counts the diagonal entry n_{ii} twice. If there is no association between rows and columns, the probability of the set of marginal totals (n_i), given the sample size N, is

$$\sum \frac{N!}{\prod n_{i.}!} \frac{N!}{\prod n_{.i}!} \prod p_i^{2n_i}$$

where the summation is over all values of $n_{i.}$ and $n_{.i}$ for which $n_{i.} + n_{.i} = n_i$ $(i = 1, 2, \cdots, t)$, and where p_i is the probability of an entry in row i. This sum is equal to the coefficient of $x_1^{n_1} \cdots x_t^{n_t}$ in

$$\{(p_1 x_1 + \cdots + p_t x_t)^N (p_1 x_1 + \cdots + p_t x_t)^N\}$$

and so is equal to

$$\frac{(2N)!}{\prod n_i!} \prod p_i^{n_i} \tag{C.1}$$

[†] See Section 6.2.

Formula 6.11 follows readily from this. A similar argument could be used to calculate the probability of getting a set of values of $n_i = n_{i..} + n_{.i.} + n_{..i}$ in a three-dimensional cubical contingency table when the probabilities of ith rows, columns, and shafts are equal ($i = 1, 2, \cdots, t$) and the rows, columns and shafts are statistically independent. The conditional probability of the interior of the table could then be obtained, conditional on the values of the n_i's.

Appendix D.   A Minimax Property of Singular Values of a Matrix[†]

There is a well-known minimax property of quadratic forms (Courant and Hilbert,[13] p. 31; and Bellman,[4] pp. 113–118, where several applications are given) to the following effect:

Let B be a symmetric matrix,[‡] and consider the maximum of the quadratic form $x'Bx$, where x is of unit length and is subjected to p homogeneous linear constraints. Then take the minimum of this maximum, when the homogeneous linear constraints are allowed to vary. This "minimax" is equal to the pth eigenvalue of B. It is natural to ask whether there is an analogue of this theorem for bilinear forms. Our purpose in this appendix is to prove such a theorem. (Note the corollary at the end of this appendix, also.)

THEOREM D.1

Consider the maximum of the bilinear form $x'Ay$, when x and y are of unit length, and are respectively constrained by p and q homogeneous linear constraints. Take the minimum of this maximum when the $p + q$ constraints are allowed to vary arbitrarily. This minimax is equal to κ_{p+q+1} the $(p+q+1)$-st singular value of A.

In this statement we can of course replace the bilinear form by $x'Ay/(\,|\,x\,|\cdot|\,y\,|)$, in which case we do not need to specify that x and y are of unit lengths.

† See Section 7.4.
‡ In this appendix we do not denote vectors or matrices by boldface type.

The proof given, for example, by Courant and Hilbert,[13] pp. 32–34, for the corresponding property of symmetric matrices and quadratic forms, does not seem to be readily adaptable to unsymmetric matrices and bilinear forms, but a proof can be constructed along different lines. We start with a lemma.

LEMMA

Let B be a symmetric non-negative definite n × n matrix and let P be a square n × n matrix of nullity at most p. Then

$$\min_{P} (q\text{th eigenvalue of } PBP') = (p + q)\text{th eigenvalue of } B$$

where we define the rth eigenvalue of B as zero when r > n.

Proof of Lemma:

Denote real Euclidean n-dimensional space by R_n. We consider that a definite orthogonal coordinate system has been selected, so that linear transformations of R_n, and matrices of n columns can be identified. While discussing the lemma, we shall be concerned only with matrices that are either vectors or have n rows and n columns, and our linear transformations will correspond to these square matrices. If Q is a linear transformation, we denote its nullity by $\mathscr{N}Q$. We use a vertical stroke to mean "given". We have

$$\min_{P} (q\text{th eigenvalue of } PBP' \mid \mathscr{N}P \le p)$$
$$= \min_{P} \{\min_{Q} [\max_{x} (x'PBP'x \mid x'x = 1, x \in QR_n) \mid \mathscr{N}Q \le q] \mid \mathscr{N}P \le p\}$$

where it is legitimate to restrict Q to be an orthogonal projection.† This follows from the minimax property for quadratic forms. Now let $x = Q\xi$, and we get

$$\min_{P} \{\min_{Q} [\max_{\xi} (\xi'Q'PBP'Q\xi \mid \xi'\xi = 1, \xi \in R_n) \mid \mathscr{N}Q \le q] \mid \mathscr{N}P \le p\}$$

where the (geometrical) justification for writing $\xi'\xi = 1$ is that the maximum with respect to ξ is assumed when $\xi \in QR_n$. Now the class of vectors $P'Q\xi$ (where P is a linear operator, Q is an orthogonal projection operator, $\mathscr{N}P \le p$, $\mathscr{N}Q \le q$, $\xi \in R_n$) is identical with the class of vectors $S\xi$, where $\xi \in R_n$ and S is an orthogonal projection operator with $\mathscr{N}S \le p + q$. So our expression equals

$$\min_{S} \{\max_{\xi} (\xi'S'BS\xi \mid \xi'\xi = 1, \xi \in SR_n) \mid \mathscr{N}S \le p + q\}$$

since the maximum when ξ ranges over R_n is achieved when it ranges only over SR_n. Now write $\eta = S\xi$, which is equal to ξ when $\xi \in SR_n$.

† It should be noted that the matrix of an orthogonal projection is not generally orthogonal.

Our expression is equal to

$$\min_{S} \{\max_{\eta} (\eta' B \eta \mid \eta' \eta = 1, \eta \in SR_n \mid \mathcal{N}S \leq p + q\}$$
$$= (p + q + 1)\text{st eigenvalue of } B$$

again by the minimax property for quadratic forms. This proves the lemma.

Proof of Theorem D.1:
    Let

$$\mu(P, Q) = \max_{x,y} (x'Ay \mid x'x = y'y = 1, x \in PR_m, y \in QR_n)$$

so that we wish to prove that

$$\min_{P,Q} \{\mu(P, Q) \mid \mathcal{N}P \leq p, \mathcal{N}Q \leq q\} = \kappa_{p+q+1}$$

The set of vectors of the form PR_m, where P is a linear operator of specified nullity, is the same as the set when P is restricted to being an orthogonal projection, which we shall then suppose. We first consider the maximum of $x'Ay(x'x = 1, x \in PR_m)$ with y held fixed. Since $x'Ay$ is the scalar product of x with Ay, its maximum is equal to the length of the orthogonal projection of the vector Ay into the space PR_m; that is, it is equal to $\mid PAy \mid$. Hence

$$(\mu(P, Q))^2 = \max_{y} (y'A'P'PAy \mid y'y = 1, y \in QR_n)$$

Therefore

$$\min_{Q} \{(\mu(P, Q))^2 \mid \mathcal{N}Q \leq q\} = q\text{th eigenvalue of } A'P'PA$$
$$\text{(by the minimax property for quadratic forms)}$$
$$= q\text{th eigenvalue of } PAA'P$$

since the eigenvalues of the Gram matrix CC', for any matrix C, are equal to those of $C'C$, apart from zeros. The theorem now follows immediately from the lemma.

COROLLARY
    Suppose that one row (or column) of a matrix is deleted or replaced by zeros, and let the singular values of the new matrix be $\kappa_1' \geq \kappa_2' \geq \cdots$. Then

$$\kappa_1 \geq \kappa_1' \geq \kappa_2 \geq \kappa_2' \geq \cdots \tag{D.1}$$

The corresponding statement for more deletions is immediately deducible from this special case, and so we do not state it separately.

Appendix E.   Learning from Experience in the Treatment of Contingency Tables†

We asserted in Chapter 6, page 51, that, in a Bayesian treatment of contingency tables it is necessary to learn from experience in the sense that the relative frequencies within one row affect the initial Type II distributions in the other rows. In this appendix we shall show that if a credibilist ignores this principle he is driven into a contradiction. In effect he would be assuming W. E. Johnson's sufficiency postulate for the four-category multinomial distribution, although it is not clear that Johnson would have supported this postulate when there is a *natural* expression of the four categories as a 2 × 2 array. The categories can of course always be expressed as a 2 × 2 array but not necessarily in a natural manner. I know no logical definition of "natural": often the expression as a 2 × 2 array will be possible with little artificiality. For example, if the names of the categories are 1, 2, 3, 4, then we could put them in a 2 × 2 array whose rows were labeled "small" and "large" and columns labeled "odd" and "even." Whatever the logical difficulties, there is no doubt that the credibilist is often forced to distinguish between natural and artificial classifications, where "natural'' must be interpreted *relative to the matter under investigation*. The judgment of naturalness has not been formalized and is therefore subjective.

† See Section 4.2 and Chapter 6.

We mentioned without proof in Chapter 8 that Johnson's sufficiency postulate is not always reasonable, in particular when the "frequencies of the frequencies" can be smoothed. The objection given in this appendix is entirely different and has the added mathematical interest of showing that the solution of a nonlinear integral equation can shed light on the philosophy of statistics.

Let p_1, p_2, p_3, p_4 be , in reading order, the four physical probabilities in a 2×2 contingency table. We shall suppose that the contingency table is obtained by random sampling, so that the four observed frequencies form a four-category multinomial sample, and also the row totals form a binomial sample, as do the column totals. The physical probabilities corresponding to the two rows are $p_1 + p_2$ and $p_3 + p_4$, and those corresponding to the two columns are $p_1 + p_3$ and $p_2 + p_4$. Moreover each row and each column also forms a binomial sample.

The physical probabilities p_1, p_2, p_3, and p_4 can be regarded as the second-generation probabilities of two distinct dichotomous trees: in one tree the first-generation probabilities are $p_1 + p_2$ and $p_3 + p_4$, and in the other they are $p_1 + p_3$ and $p_2 + p_4$. (Compare Chapter 4, page 30.) A credibilist would like to be able to assume the same initial Type II distribution for the parameter p of all six binomial populations, even though p_1, for example, corresponds to an object, junior by one generation in the classification of all objects in the world, than does $p_1 + p_2$. It would be impracticable for a credibilist to allow for so comprehensive a classification, although the subjectivist might do so to some extent, either consciously or unconsciously.

Suppose that the credibilist's Type II initial distribution of a binomial physical probability p has the density $f(p)$. We could write $dF(p)$ for a probability element, but we shall allow Dirac functions, so there is virtually no loss of generality in talking about a density function. (The following argument would go through just as well with Stieltjes integrals.) This density function will apply to $p_1 + p_2, p_3 + p_4, p_1 + p_3$, and $p_2 + p_4$, and also to $p_1/(p_1 + p_2)$ given $p_1 + p_2$, and so forth. If we write P.D. for "probability density," we have

$$\text{P.D.}(p_1 = x \mid p_1 + p_2 = y) = \frac{f(x/y)}{y}$$

Moreover

$$\text{P.D.}(p_1 + p_3 = x)$$
$$= \int \text{P.D.}(p_1 + p_2 = y) \, dy \int \text{P.D.}(p_1 = z \mid p_1 + p_2 = y)$$
$$\times \text{P.D.}\left(p_3 = x - z \mid p_3 + p_4 = 1 - y, \frac{p_1}{p_1 + p_2} = \frac{z}{y}\right) dz \quad \text{(E.1)}$$

Suppose now that the credibilist ignores the principle mentioned in the first paragraph of this appendix. Then we obtain the nonlinear integral equation

$$f(x) = f(1 - x) = \int_0^x dx \int_z^{1-x+z} \frac{1}{y(1-y)} f(y) f\left(\frac{z}{y}\right) f\left(\frac{x-z}{1-y}\right) dy \quad \text{(E.2)}$$

where, for all x,

$$f(x) = f(1 - x), \qquad f(x) \geq 0$$
$$f(x) = 0, \text{ whenever } x < 0 \text{ or } x > 1$$

and

$$\int_0^1 f(x) \, dx = 1$$

In the solution of this equation we shall make use of the delta function $\delta(x)$, defined formally by the identity

$$\int_{-\infty}^{\infty} g(x)\delta(x) \, dx = g(0)$$

for some class of smooth functions $g(x)$. We shall also make use of the "sliced" delta function $\delta_1(x)$, defined by

$$\int_0^{\infty} g(x)\delta_1(x) \, dx = \frac{1}{2}g(0)$$

THEOREM E.1

1. *There are no continuous integrable solutions of Equation E.2;*
2. *The only integrable solutions are*

$$f(x) = \delta(x - \tfrac{1}{2}) = f_0(x) \quad \text{(E.3)}$$

say, and

$$f(x) = \delta_1(x) + \delta_1(1 - x) = f_1(x) \quad \text{(E.4)}$$

Proof:

Denote the right side of Equation E.2 by $Tf(x)$ and let

$$\xi = \frac{x - z}{1 - y}$$
$$\eta = y$$
$$\zeta = \frac{z}{y}$$

Then, for any bounded measurable function $g(x)$, we have

$$\int_0^1 g(x)Tf(x) \, dx = \int_0^1 \int_0^1 \int_0^1 g(\xi + \eta\zeta - \eta\xi)f(\xi)f(\eta)f(\zeta) \, d\xi \, d\eta \, d\zeta$$

$$\text{(E.5)}$$

First take $g(x)$ identically equal to 1 in Equation E.5. We deduce that if $f(x)$ is a density function, then so is $Tf(x)$. It can further be seen directly from Equation E.2 that symmetry $f(x) = f(1 - x)$ is preserved under the transformation T. It follows that

$$\int_0^1 xTf(x)\, dx = \frac{1}{2}$$

a fact that also follows from Equation E.5 by putting $g(x) = x$. Next we consider the second moment of $Tf(x)$,

$$\int_0^1 x^2 Tf(x)\, dx = \int_0^1 \int_0^1 \int_0^1 (\xi + \eta\zeta - \eta\xi)^2 f(\xi)f(\eta)f(\zeta)\, d\xi\, d\eta\, d\zeta$$
$$= b + 2b^2 + 2a^3 - 2ab - 2a^2 b$$

where $a = 1/2$, and b, are the first and second moments of $f(x)$ about the origin. But if $f(x)$ is a solution of Equation E.2, then $Tf(x) = f(x)$, and we deduce that

$$b = b + 2b^2 + \tfrac{1}{4} - b - \tfrac{1}{2}b$$

so that $b = 1/4$ or $1/2$, the variance of $f(x)$ is 0 or $1/4$, and therefore $f(x)$ is $f_0(x)$ or $f_1(x)$. Also, it is straightforward to verify that these are both solutions of Equation E.2, since they reduce the right side of Equation E.5 to

$$\int_0^1 g(x)f(x)\, dx$$

This proves the theorem.

The density f_0 corresponds to the "null" hypothesis H_0, that $p = 1/2$; the density f_1 to the hypothesis that, with Type II probability 0.5, $p = 0$ and otherwise $p = 1$. For the purpose of testing the null hypothesis, H_1 is a hopelessly inadequate model of the alternative. For the factor in favor of H_0 provided by any binomial sample, consisting of r successes and s failures, would be infinite if H_1 were the alternative hypothesis and if r and s were both positive.

Matters are not quite so bad for the purpose of *estimating* p once the null hypothesis has been rejected, since, in a sense to be described, it is possible to make use of the hypothesis H_1. Given a binomial sample consisting of r successes and s failures, the final density of p will be

$$\frac{p^r(1 - p)^s f(p)}{\int_0^1 x^r(1 - x)^s f(x)\, dx} \tag{E.6}$$

where $f(p)$ is the initial density. If we put $f(p) = f_1(p)$, then Equation E.6 reduces to the indeterminate form 0/0 whenever r and s are both positive, and $p \neq 0, p \neq 1$. This is not surprising, since the likelihood of H_1 is 0 in this case. But H_1 can be slightly modified so as to overcome this difficulty. Let $h(x)$ denote the Haldane improper density $h(x) \propto x^{-1}(1 - x)^{-1}$, which we here interpret as something like a "generalized function,"

$$h(x) = h_k(x) = \frac{\Gamma(2k)}{\Gamma(k)^2} x^{k-1}(1 - x)^{k-1}$$

where k is negligibly small and positive. Nearly all the area under the graph of $h(x)$ is crowded at the points $x = 0$ and $x = 1$. So, although $h(x)$ is not quite the same as $f_1(x)$, it does correspond to almost the same distribution function, and it will very nearly satisfy the nonlinear integral Equation E.2. For the purpose of testing the null hypothesis H_0, the density $h(x)$ will be almost as bad as $f_1(x)$, so it cannot be regarded as satisfactory. But if we replace $f(p)$ by $h(p)$, in Equation E.6, it is no longer of the indeterminate form 0/0, in fact it reduces to a beta density with parameters $r - 1$ and $s - 1$, to a degree of approximation as good as one wishes.

Note that $h(x)$ is by no means unique, as an approximation to $f_1(x)$, since we can replace $h_k(x)$ by any function of the form

$$\frac{j(x)x^{k-1}(1 - x)^{k-1}}{\int_0^1 j(x)x^{k-1}(1 - x)^{k-1} \, dx}$$

where $j(x)$ is symmetrical, strictly positive, and continuous in the closed interval $[0, 1]$. When k is small, we get an approximation to $f_1(x)$ in the same sense as before, but we get different final distributions. The Bayes-Laplace uniform distribution and the Jeffreys-Perks distribution are not of this form.

The following remarks concerning moments are of some mathematical interest.

The complete set of moments of f_0 and f_1 are

$$\int_0^1 x^n f_0(x) \, dx = 2^{-n} \qquad (n = 0, 1, 2, \cdots)$$

$$\int_0^1 x^n f_1(x) \, dx = 1 \qquad (n = 0)$$

$$= \tfrac{1}{2} \qquad (n = 1, 2, 3, \cdots)$$

For both functions, the moments for $n = 0, 1$, and 2 determine all the rest.

One can verify directly from Equation E.5 that all the moments of $Tf_0(x)$ are the same as those of $f_0(x)$. For if the nth moment of $f_0(x)$ is denoted by c_n, we see that the nth moment of $Tf_0(x)$ is

$$n! \sum_{\lambda, \mu, \nu} \frac{(-1)^\nu c_{\lambda + \nu} c_{\mu + \nu} c_\mu}{\lambda! \mu! \nu!} = (\tfrac{1}{2} + \tfrac{1}{4} - \tfrac{1}{4})^n = 2^{-n} \qquad (\lambda + \mu + \nu = n)$$

Similarly, but with a little more trouble, one can verify directly from Equation E.5 that the moments of $Tf_1(x)$ are the same as those of $f_1(x)$.

The moments of the final distribution can be obtained by multiplying Equation E.6 by p^n and integrating. The reader might find it interesting to check what happens when $f(x)$ is replaced by $f_0(x)$, $f_1(x)$, and $h(x)$ in turn.

E.1 Generalization to $m \times n$ Contingency Tables

Although the earlier discussion of the 2×2 contingency table is sufficient for our purposes, it is interesting to give another proof that applies equally well to an $m \times n$ contingency table, especially as it is simpler. The previous proof is included because it exhibits an interesting example of a problem in probability leading to a nonlinear integral equation, and because the method of solution of the integral equation, by the use of moments, might have other applications. Also, under a variety of modifications of Johnson's sufficiency postulate, one would have to make use of the more general integral Equation E.1.

Given the row-total probabilities $p_{i.}$, let the conditional probabilities in the individual cells be q_{ij} ($i = 1, 2, \cdots, m$; $j = 1, 2, \cdots, n$). Then the column totals are

$$p_{.j} = \sum {}_{i} p_{i.} q_{ij}$$

We now assume that each of the m vectors, consisting of the conditional probabilities in a row, are independently sampled with identical Type II distributions, and also that the vector $(p_{.j})$ has this same distribution. We do not need to assume that this distribution is symmetrical in the n components, and we assume nothing about the distribution of the row totals. We write $\mathscr{E}_{II}(q_{ij}) = b_j$. Further, we write $c_{jj'}$ for the covariance

$$c_{jj'} = \mathscr{E}_{II}(q_{ij} - b_j)(q_{ij'} - b_{j'})$$

We have

$$c_{jj'} = \mathscr{E}_{II}(p_{.j} - b_j)(p_{.j'} - b_{j'})$$
$$= \mathscr{E}_{II} \sum {}_{i,i'} p_{i.} p_{i'.} \mathscr{E}_{II}(q_{ij} - b_j)(q_{i'j'} - b_{j'})$$

as is easily seen. Terms having $i \neq i'$ vanish since the (Type II) sampling of the rows was independent, and we obtain

$$c_{jj'} = c_{jj'} \mathscr{E}_{II} \sum p_{i.}^2$$

If $\mathscr{E}_{II} \sum p_{i.}^2 = 1$, then almost certainly one of the $p_{i.}$'s is unity and the rest zero; that is, only one of the rows of the contingency table actually occurs. Otherwise $c_{jj'} = 0$ for all j and j', and then it is almost certain that $q_{ij} = b_j$ for all j.

References

1. Allais, Maurice, *Fondements d'une Théorie Positive des Choix Comportant un Risque et Critique des Postulats et Axiomes de l'Ecole Américaine Econométrie*, Centre de la Recherche Scientifique, Paris 257–332 (1953).
2. Bartlett, M. S., "Probability and Chance in the Theory of Statistics," *Proc. Roy. Soc. (London)*, A, **141**, 518–534 (1933).
3. Bartlett, M. S., "Contingency Table Interactions," *J. Roy. Statist. Soc.*, Suppl., **2**, 248–252 (1935).
4. Bellman, Richard, *Introduction to Matrix Analysis*, McGraw-Hill, New York, 1960.
5. Belonogov, G. G., "On Some Statistical Regularities in Written Russian," *Vopr. Jazykoznanija*, **7**, 100–101 (1962). (In Russian.)
6. Boltzmann, L., *Wiener Sitzungsberichte*, **76**, 373 (1877). Cited by P. and T. Ehrenfest in *The Conceptual Foundations of the Statistical Approach in Mechanics* (Eng. trans. by M. J. Maravcsik), Cornell University Press, Ithaca, N.Y., 1959, pp. 27 and 29.
7. Broad, C. D., "On the Relation Between Induction and Probability," *Mind*, **27**, 389–404 (1918); **29**, 11–45 (1920).
8. Brown, David T., "A Note on Approximations to Discrete Probability Distributions," *Information and Control*, **2**, 386–392 (1959).
9. Carnap, Rudolf, *The Continuum of Inductive Methods*, University of Chicago Press, Chicago, Ill., 1952.
10. Champernowne, D. G., "A Model of Income Distribution," *Econ. J.*, **63**, 318–351 (1953).
11. Cochran, William G., "The χ^2 Correction for Continuity," *Iowa State College J. Sci.*, **16**, 421–436 (1942).
12. Cohen, John, *Chance, Skill, and Luck*, Pelican, London, 1960.
13. Courant, Richard and David Hilbert, *Methods of Mathematical Physics, I*, Interscience Publishers, New York 1953, 1962.

14. Darroch, J. N., "Interactions in Multifactor Contingency Tables," *J. Roy. Statist. Soc.*, B, **24**, 251–263 (1962).

15. Davidson, Donald, Patrick Suppes, and Sidney Siegal, *Decision Making: An Experimental Approach*, Stanford University Press, Stanford, Calif., 1957.

16. Deming, W. Edwards, and Frederick W. Stephan, "On a Least Squares Adjustment of a Sampled Frequency Table," *Ann. Math. Statist.*, **11**, 427–444 (1940).

17. Duncan, W. J., R. A. Frazer, and A. R. Collar, *Elementary Matrices and Some Applications to Dynamics and Differential Equations*, Cambridge University Press, 1957.

18. Eck. Richard V., "Non-randomness in Amino-Acid 'Alleles,'" *Nature*, **191**, 1284-1285 (1961).

19. Edgeworth, F. Y., "Probability," *Ency. Brit. 11th ed.*, **22**, 376–403 (1910).

20. Erdélyi, Artur, Wilhelm Magnus, Fritz Oberhettinger, and Francesco G. Tricomi, *Tables of Integral Transforms, I*, McGraw-Hill, New York, 1954.

21. De Finetti, Bruno, "La prévision, ses lois logiques, ses sources subjectives," *Ann. Inst. Henri Poincaré*, **7**, 1-68 (1937).

22. Fisher, Ronald A., *Statistical Methods and Scientific Inference*, Oliver and Boyd, Edinburgh, 1956.

23. Fisher, Ronald A., *Statistical Methods for Research Workers*, Hafner, New York, 13th ed., 1958.

24. Fréchet, Maurice, *Les probabilités associées à un système d'événements compatibles et dépendents, II*, Hermann et Cie, Paris, 1943, Chapter III.

25. Freedman, David A., "Invariants Under Mixing Which Generalize de Finetti's Theorem: Continuous Time Parameter," *Ann. Math. Statist.*, **34**, 1194–1216 (1963).

26. Gibbs, J. W., *The Scientific Papers of J. Willard Gibbs*, **1**, Dover Publications, New York, 1961.

27. Good, I. J., *Probability and the Weighing of Evidence*, Griffin, London, 1950; Hafner, New York, 1950.

28. Good, I. J., "Rational Decisions," *J. Roy. Statist. Soc.*, B, **14**, 107–114 (1952).

29. Good, I. J., "On the Population Frequencies of Species and the Estimation of Population Parameters," *Biometrika*, **40**, 237–264 (1953).

30. Good, I. J., "The Appropriate Mathematical Tools for Describing and Measuring Uncertainty," *Uncertainty and Business Decisions*, ed. by C. F. Carter, G. P. Meredith, and G. L. S. Shackle; University Press, Liverpool, 1954, pp. 19–34.

31. Good, I. J., "On the Estimation of Small Frequencies in Contingency Tables," *J. Roy. Statist. Soc.*, B, **18**, 113–124 (1956).

32. Good, I. J., "Distribution of Word Frequencies," *Nature*, **179**, 595 (1957).

33. Good, I. J., "Saddlepoint Methods for the Multinomial Distribution," *Annals Math. Statist.*, **28**, 861–881 (1957). Cont. in **32**, 535–548 (1961).

34. Good, I. J., "Significance Tests in Parallel and in Series," *J. Amer. Statist. Assn.*, **53**, 799–813 (1958).

35. Good, I. J., "The Interaction Algorithm and Practical Fourier Analysis," *J. Roy. Statist. Soc.*, B, **20**, 361–372 (1958); **22**, 372–375 (1960).
36. Good, I. J., "Kinds of Probability," *Science*, **129**, 443–447 (1959). Italian translation in *L'Industria*, 1959.
37. Good, I. J., Review of H. A. Simon, Ref. 103, Rev. no. A593, *Math. Rev.*, **24** (1962).
38. Good, I. J., "How Rational Should a Manager Be?" *Management Science*, **8**, 383–393 (1962).
39. Good, I. J., "A Compromise Between Credibility and Subjective Probability," *International Congress of Mathematicians, Abstracts of Short Communications*, Stockholm, 1962. p. 160.
40. Good, I. J., Contribution to the discussion of a paper by Charles Stein, *J. Roy. Statist. Soc.*, B, **24**, 289–291 (1962).
41. Good, I. J., "Botryological Speculations," *The Scientist Speculates*, Heinemann, London, 1962; Basic Books, New York, 1963, pp. 120–132.
42. Good, I. J., "Maximum Entropy for Hypothesis Formulation, Especially for Multidimensional Contingency Tables," *Annals Math. Statist.*, **34**, 911–934 (1963).
43. Good, I. J., "A Categorization of Classification," to be published in the proceedings of a conference on mathematics and computer science in biology and medicine at Oxford, 1964, by the British Medical Research Council.
44. Good, I. J., and G. H. Toulmin, "The Number of New Species, and the Increase of Population Coverage, When a Sample is Increased," *Biometrika*, **43**, 45–63 (1956).
45. Goodman, Leo A., "Interactions in Multidimensional Contingency Tables," *Annals Math. Statist.*, **35**, 632–646 (1965).
46. Haag, Jules, "Sur un problème général de probabilités et ses diverses applications," *Proc. International Congress of Mathematics, Toronto, 1924*, Toronto, 1928, 659–674.
47. Haldane, J. B. S., "On a Method of Estimating Frequencies," *Biometrika*, **33**, 222–225 (1945).
48. Haldane, J. B. S., "The Precision of Observed Values of Small Frequencies," *Biometrika*, **35**, 297–300 (1948).
49. Halmos, Paul R., *Finite Dimensional Vector Spaces*, Van Nostrand, Princeton, 1958.
50. Hardy, G. F., in correspondence in Insurance Record, 1889, reprinted in *Trans. Fac. Actuaries*, **8**, (1920) (mentioned in Perks, Ref. 93).
51. Hardy, G. H., *Divergent Series*, Oxford University Press, 1949.
52. Harman, Harry H., *Modern Factor Analysis*, University of Chicago Press, 1960.
53. Harris, Bernard, "Determining Bounds on Integrals with Applications to Cataloguing Problems," *Annals Math. Statist.*, **30**, 521–548 (1959).
54. Hartmanis, Juris, "The Application of Some Basic Inequalities for Entropy," *Information and Control*, **2**, 199–213 (1959).
55. Herdan, G., *Type-Token Mathematics*, Mouton, 's-Gravenhage, 1960.
56. Hewitt, Edwin, and Leonard J. Savage, "Symmetric Measures on Cartesian Products," *Trans. Amer. Math. Soc.*, **80**, 470–501 (1955).

57. Hildebrandt, T. H., and I. J. Schoenberg, "On Linear Functional Operations and the Moment Problem for a Finite Interval in One or Several Dimensions," *Annals Math.*, **34**, 317–328 (1933).

58. Hinčin, A. Ya, "O klassah ekvivalentnyh sobitiĭ," *Dokl. Akad. Nauk SSSR*, **85**, 713–714 (1952).

59. Hume, David, *An Enquiry Concerning Human Understanding*, London, 1748.

60. Ishii, Goro, "Intraclass Contingency Tables," *Ann. Inst. Statist. Math.*, **12**, 161–207 (1960).

61. Jaynes, E. T., "Information Theory and Statistical Mechanics," *Phys. Rev.*, **106**, 620-630 (1957); **108**, 171–190 (1957).

62. Jaynes, E. T., "New Engineering Applications of Information Theory," in *Proc. First Symp. Engineering Applications of Function Theory and Probability*, ed. by J. L. Bogdanoff and F. Kozin, Wiley, New York and London, 1963, pp. 163-203.

63. Jeffreys, Harold, *Theory of Probability*, Clarendon Press, Oxford, 3rd ed., 1961.

64. Jeffreys, Harold, "An Invariant Form for the Prior Probability in Estimation Problems," *Proc. Roy. Soc. (London)*, A, **186**, 453–461 (1946).

65. Johnson, N. L., "An Approximation to the Multinomial Distribution: Some Properties and Applications," *Biometrika*, **47**, 93–102 (1960).

66. Johnson, William Ernest, *Logic, Part III. The Logical Foundations of Science*, Cambridge University Press, 1924, Appendix on Education, pp. 178–189 (esp. p. 183).

67. Johnson, W. E., Appendix (ed. by R. B. Braithwaite) to "Probability: Deductive and Inductive Problems," *Mind*, **41**, 421–423 (1932).

68. Jones, K. Spärck, "Mechanized Semantic Classification," *Proc. First International Conf. on Mechanized Translation*, National Physical Laboratory, Teddington, England, 1963.

69. Kemble, Edwin C., "Is the Frequency Theory of Probability Adequate for All Scientific Purposes?" *Am. J. Phys.*, **10**, 6-16 (1942).

70. Kendall, Maurice G., *Advanced Theory of Statistics*, Vol. 1, Charles Griffin, London, 1945.

71. Keynes, John Maynard, *A Treatise on Probability*, Macmillan, London and New York, 1921.

72. Keynes, John Maynard, *Essays in Biography*, "F. P. Ramsey, 1903-1930," Rupert Hart Davis, London, 1933, 1951, pp. 239–252 (especially p. 243).

73. Kincaid, W. M., "The Combination of Tests Based on Discrete Distributions," *J. Amer. Statist. Assn.*, **58**, 10–19 (1963).

74. Koopman, Bernard Osgood, "The Axioms and Algebra of Intuitive Probability," *Annals Math.*, **41**, 269–292 (1940).

75. Koopman, Bernard Osgood, "The Bases of Probability," *Bull. Amer. Math. Soc.*, **46**, 763–774 (1940).

76. Kullback, S., M. Kupperman, and H. H. Ku, "An Application of Information Theory to Contingency Tables," *J. Res. Nat. Bur. Std.*, B, **66B**, 217–243 (1962).

77. Lange, J., *Crime and Destiny*, Allen and Unwin, London, 1931.

78. Laplace, Pierre Simon, "Mémoire sur la probabilité des causes par les événements," *Mém. de. l'Acad. R. de. Sci. Paris*, 6, 621–656 (1774).
79. Lewis, P. M., II, "Approximating Probability Distributions to Reduce Storage Requirements," *Information and Control*, 2, 214–225 (1959).
80. Lidstone, G. J., "Note on the General Case of the Bayes-Laplace Formula for Inductive or *a posteriori* Probabilities," *Trans. Fac. Actuar.*, 8, 182–192 (1920).
81. Lindley, D. V., "Fiducial Distributions and Bayes' Theorem," *J. Roy. Statist. Soc.*, B, 20, 102–107 (1958).
82. Mandelbrot, Benoît, *Information Theory: Third London Symposium*, ed. by Colin Cherry, Butterworths, London, 1956, p. 135.
83. Mauldon, J. G., "A Generalization of the Beta-Distribution," *Ann. Math. Statist.*, 30, 509–520 (1959).
84. Maxwell, A. E., *Analysing Qualitative Data*, Methuen, London; Wiley, New York, 1961, Ch. 6, "The Analysis of 2^n Contingency Tables."
85. von Mises, R., "Über Aufteilungs- und Besetzungs-Wahrscheinlichkeiten," *Acta [Trudy] Univ. Asiae Mediae, Ser.* V-a, *Fasc.*, 27, 21 (1939).
86. Mood, Alexander McFarlane, *Introduction to the Theory of Statistics*, McGraw-Hill, New York, 1950.
87. Mosimann, J. E., "On the Compound Multinomial Distribution, the Multivariate β-Distribution, and Correlations Among Proportions," *Biometrika*, 49, 65–82 (1962).
88. Mosimann, J. E., "On the Compound Negative Multinomial Distribution and Correlations Among Inversely Sampled Pollen Counts," *Biometrika*, 50, 47–54 (1963).
89. von Neumann, John, and Oskar Morgenstern, *The Theory of Games and Economic Behavior*, 2nd ed., Princeton University Press, Princeton, N.J., 1947, appendix.
90. Pearson, E. S., "Note on Professor Haldane's Paper Regarding the Treatment of Rare Events," *Biometrika*, 35, 301–303 (1948).
91. Pearson, E. S., "On Questions Raised by the Combination of Tests Based on Discontinuous Distributions," *Biometrika*, 37, 383–398 (1950).
92. Pearson, Karl, *Tables of the Incomplete Γ Function*, Cambridge University Press, 1957.
93. Perks, Wilfred, "Some Observations on Inverse Probability Including a New Indifference Rule," *J. Inst. Actuar.*, 73, 285–312 (1947).
94. Poisson, S. D., *Calcul des Jugements*, Paris (1837).
95. Ramsey, Frank Plumpton, *Foundations of Mathematics*, Kegan Paul, London; Harcourt, New York, 1931 Chs. 7 and 8.
96. Robbins, Herbert E., "An Empirical Bayes Approach to Statistics," *Proc. of the Third Berkeley Symp. on Math. Statist. and Probability*, University of California Press, Berkeley, Calif. 1956, pp. 157–163.
97. Robbins, Herbert, "The Empirical Bayes Approach to Statistical Decision Problems," *Ann. Math. Statist.*, 35, 1–20 (1964).
98. Roy, A. D., "Some Notes on Pistimetric Inference," *J. Roy. Statist. Soc.*, B, 22, 338–347 (1960).
99. Russell, Bertrand, *Human Knowledge*, George Allen and Unwin, London, 1948 (esp. p. 359).

100. Savage, Leonard J., *The Foundations of Statistics*, Wiley, New York and London, 1954.
101. Shanks, Daniel, "Nonlinear Transformations of Divergent Series and Slowly Convergent Series," *J. Math. Phys.*, **34**, 1–42 (1955).
102. Shannon, Claude E., and W. Weaver, *The Mathematical Theory of Communication*, University of Illinois Press, Urbana, Ill., 1949.
103. Simon, Herbert A., "Some Further Notes on a Class of Skew Distribution Functions," *Information and Control*, **3**, 80–88 (1960).
104. Smith, C. A. B., "Consistency in Statistical Inference and Decision," *J. Roy. Statist. Soc.*, Ser. **B**, **23**, 1–25 (1961).
105. Smithies, F., *Integral Equations*, Cambridge University Press, 1958.
106. Thatcher, A. R., "Relationships Between Bayesian and Confidence Limits for Predictions," *J. Roy. Statist. Soc.*, B, **26**, 176–210, with discussion (1964).
107. Thorndike, E. L., and I. Lorge, *The Teacher's Word Book of 30,000 Words*, Columbia University Press, New York, 1954.
108. Wagner, Karl Willi, "Theorie der dielektrischen Nachwirkung," *Elektrotech. Z.*, **45**, 1279–1281 (1913).
109. Weyl, Hermann, "Inequalities Between the Two Kinds of Eigenvalues of a Linear Transformation," *Proc. Nat. Acad. Sci. U.S.*, **35** (1949).
110. Whittaker, Edmund T., and G. N. Watson, *A Course of Modern Analysis*, 4th ed., Cambridge University Press, 1935.
111. Whittle, P., "On Principal Components and Least Square Methods of Factor Analysis," *Skand. Aktuarietidskr.*, **35**, 223–239 (1952).
112. Widder, David Vernon, *The Laplace Transform*, Princeton University Press, 1941.
113. Wilks, S. S., "The Large-Sample Distribution of the Likelihood Ratio for Testing Composite Hypotheses," *Ann. Math. Statist.*, **9**, 60-62 (1938).
114. Wilks, S. S., *Mathematical Statistics*, Wiley, New York, 1962.
115. Zipf, G. K., *Human Behavior and the Principle of Least Effort*, Addison-Wesley, Cambridge, Mass., 1949.

Index

Printed in the United States
by Baker & Taylor Publisher Services